EINSTEIN

HIS LIFE AND TIMES

EINSTEIN

HIS LIFE AND TIMES

PHILIPP FRANK

TRANSLATED FROM A GERMAN MANUSCRIPT BY

GEORGE ROSEN

EDITED AND REVISED BY SHUICHI KUSAKA

DA CAPO PRESS

Cataloging-in-Publication data for this book is available from the Library of Congress.
ISBN 0–306–81109–X

Second Da Capo Press edition 2002
This Da Capo Press paperback edition of *Einstein: His Life and Times* is an unabridged republication of the edition published in New York in 1947, then revised in 1953. It is reprinted by arrangement with Alfred A. Knopf, Inc.

Published by Da Capo Press
A Member of the Perseus Books Group
http://www.dacapopress.com

Da Capo Press books are available at special discounts for bulk purchases in the U.S. by corporations, institutions, and other organizations. For more information, please contact the Special Markets Department at the Perseus Books Group, 11 Cambridge Center, Cambridge, MA 02142, or call (800) 255-1514 or (617) 252-5298, or e-mail j.mccrary@perseusbooks.com.

The most incomprehensible thing about the world is that it is comprehensible.

— ALBERT EINSTEIN

ACKNOWLEDGMENTS

THE PHOTOGRAPHS reproduced in this book were obtained with the friendly help of Miss Helen Dukas of Princeton, Professor Rudolph W. Ladenburg of Princeton University, Professor Harlow Shapley of Harvard University, and Dr. and Mrs. Gustav Bucky of New York. The diagrams were designed by Mr. Gerald Holton of Harvard University, and the Index compiled with the co-operation of Miss Martha Henderson of Cambridge, Massachusetts.

CONTENTS

Contents

Contents

VII. EINSTEIN AS A PUBLIC FIGURE

VIII. TRAVELS THROUGH EUROPE, AMERICA, AND ASIA

IX. DEVELOPMENT OF ATOMIC PHYSICS

X. POLITICAL TURMOIL IN GERMANY

Contents

XI. EINSTEIN'S THEORIES AS POLITICAL WEAPONS AND TARGETS

XII. EINSTEIN IN THE UNITED STATES

ILLUSTRATIONS

INTRODUCTION

1. *"To understand Einstein" means to understand the world of the twentieth century*

       I am writing this biography of Einstein not for physicists and mathematicians, not for philosophers and theologians, not for Zionists and pacifists, but for people who want to understand something of the contradictory and complicated twentieth-century world.

It has often been said that "to understand precisely one hundredth of an inch of a blade of grass, one would have to understand the universe." But one who could achieve such understanding of a blade of grass would find nothing unclear about anything else in the universe. In a like spirit it can be said that anyone who comprehends even a little of Einstein's personality, his work, and its influence will have taken a long step toward an understanding of the world of the twentieth century.

Through a combination of fortunate circumstances I had the desire and opportunity to observe Einstein as a man and a scientist. Since my student days I had been captivated again and again by the way in which he was able to derive newly discovered, and often strange, natural phenomena from simple and elegant laws. The connection between physical and philosophic theories had also attracted me repeatedly. As time went on, one question became for me more and more an object of curiosity and often of amazement: why is it that scientific and philosophical theories that apparently have hardly anything to do with human life are so often employed to influence attitudes toward practical questions in politics and religion?

In 1912 I became Einstein's successor as professor of theoretical physics at the University of Prague; and in 1938, when I came to the United States, I again met Einstein, who had already been here for five years. I conceived the idea of taking advantage of this physical proximity to prepare an account of his life and work. When I told Einstein about this plan he said:

"How strange that you are following in my footsteps a second time!"

Before we arrived in the United States my wife often told me that I had now written so many books and papers palatable only to a small number of specialists that for a change it would be good for me to write a book more people could enjoy. As a matter of fact, I have frequently regretted the wide gap yawning between the books written for specialists in science and those for the large community of educated men and women. I had been looking for an occasion to make a contribution toward bridging this gap. I longed to write a book that could help make understandable the work done by contemporary scientists and to do so by providing more insight into the psychological and cultural background of scientific research than regular scientific books, even of the popular brand, can offer.

All these circumstances encouraged me to write this book. Many specialists have tried to dissuade me, pointing out that I would have only a choice of two evils. Either I would write to be understood by the public at large and the book would become trivial and be criticized by the scientists; or I would write it to please the specialists, but then it would be incomprehensible to others.

Such arguments did not deter me, because I did not believe there was such a fundamental difference between layman and specialist. Every specialist becomes a layman as soon as he leaves his own very narrow field. This book deals with so many fields of human life and thought that no one can be a specialist in all of them. Consequently I believe that I may, with a clear conscience, write for the laity without appearing superficial to the specialist, because in reality the complete specialist does not exist.

By training and occupation I am a mathematician and physicist, not a writer. Through this occupation one develops an aversion to exaggerations of all kinds. One acquires enthusiasm only for what is directed toward the search for truth and its presentation in a comprehensible and polished form.

In so far as pure facts are concerned, I have made partial use of earlier biographies of Einstein. The portrayal of Einstein's personality and of his position in our time, however, derives from my study of the writings of Einstein's friends and enemies, and in large measure from personal conversations with Einstein himself.

The picture of Einstein as presented throughout this book is the one I have derived from my own impressions. It is in no way Einstein's autobiography. I describe Einstein just as a scientist would describe any other remarkable, rare, and mighty natural or historical phenomenon. Only thus can justice be done to a great man.

---

2. *Einstein's popularity and incomprehensibility*

In a recent biography of one of the greatest physicists the statement occurs: "After he printed his new principles, the students on the college campus said as he passed by: there goes the man who has written a book that neither he nor anyone else understands." This appears in a biography not of Einstein, but of Isaac Newton, who in our day has so often been contrasted as an example of lucidity with the "incomprehensible" Einstein.

A contemporary of Newton extolled him in a poem that culminated in the lines:

> Nature and Nature's laws lay hid in night:
> God said, Let Newton be! and all was light.

And in our day the following lines were added:

> But not for long. The devil howling "Ho,
> Let Einstein be" restored the status quo.

This characteristic of "incomprehensibility" played a large part in the popular Einstein legend. In New York an insurance agent whom I told of my intention to write this book, astounded, said: "I hope you won't try to convince me that you can understand Einstein." When I asked why he considered it impossible, he replied categorically: "We use the word 'Einstein' as equivalent to 'incomprehensible.' When we want to say something is incomprehensible, we say 'That is Einstein.' That is why it is meaningless to say you understand Einstein."

An alleged remark by Einstein to the effect that only twelve people in the world understand his theory has become widespread. The question must be raised whether one ever actually understands anything, and whether what is called incomprehensible does not depend on the demands that one makes. Anyone who wonders that great popularity should be combined with incomprehensibility must realize that both have an important characteristic in common: the quality of being unusual.

Whatever is unusual is incomprehensible, but at the same time it possesses the power of attraction. And in the popular mind this "unusual" quality has always been linked with Einstein.

One wintry day I arrived in Princeton. The streets were piled high with snow that was already beginning to melt. There were no busses or streetcars, and I wanted to get to the residence of a mathematician who lived at some distance from the road, at 270 Mercer Street. I inquired of a man who was shoveling snow where this house might be. The man looked up from his work and, with considerable astonishment, said: "270? That's Einstein's house." As Einstein lives at 112 Mercer Street, I assured him that it was definitely not Einstein's house. "Well," he said, "we always call No. 270 Einstein's house. If you don't believe it, hop on my truck and take a look at the house." I was happy to get any vehicle in that weather and so we rode down to No. 270.

It was a house with a flat roof in the style of modern European architecture, like that of the Bauhaus, and was actually quite different from the other houses on the street, which were all more or less colonial in style. The snow-shoveler said triumphantly: "Doesn't this house look queer — very different from all its neighbors?" I could only answer: "But the house in which Einstein really lives, No. 112, looks exactly like the neighboring houses from the outside."

Many people have no idea to what branch of human knowledge the theory of relativity actually belongs. During the twenties, in Prague, I visited one of those popular lectures on Einstein's theory which were so common then. There I met a Catholic theologian with whom I was acquainted, who introduced me as a physicist to a Bishop who was present. "Oh," said the Bishop, quite amazed, "are the physicists also interested in Einstein's theory?" Later we shall see that this question was indeed strange, and yet not as inappropriate as appears at first glance.

---

3. *Superficial interpretations of Einstein's theory*

The public at large has always considered Einstein "incomprehensible." A somewhat closer examination of popular conceptions of Einstein's theories, however, reveals that there was something that people believed they understood. Obviously, anything completely incomprehensible cannot be admired. Generally, however, this nucleus of "comprehensibility" is found to be an enormous triviality. During the twenties I once arrived in a small town in Bohemia inhabited by the Sudeten Germans who were later to acquire such prominence. I arrived at the inn in the evening and found the guests smoking their pipes and drinking beer. When they learned that I was a physicist from Prague, one of them remarked he had heard that I occupied myself with Einstein's theory. He said to me: "These Einsteinian theories are not new in our town. They were known here long before Einstein. For twenty years our municipal doctor used to come to this inn, light his pipe, and take his first drink of beer with the words: 'All is relative.' Einstein did not say more."

In New York I once heard a passenger ask a bus conductor how far it was to Washington Square. The conductor replied emphatically and with some pride: "According to Einstein, 'far' is a relative idea. It depends on how much of a hurry you are in."

On another occasion I listened to a lecture by a well-known popularizer of Einstein's theory. He illustrated everything with slides. One picture showed a student in a classroom listening to the professor's boring lecture. The student looked at the clock and sighed: "This is going to go on for a long time yet. Ten more minutes — an eternity." The next picture showed the same student on a bench in the park talking with a beautiful girl. "I can stay only ten minutes longer," says the girl; and the student sighs: "Ten minutes — it will pass like a flash."

It was in 1927, however, that I had my most remarkable experience of this kind. It is of greater interest because it enables us to learn something about the role of Einstein's theories in politics. At that time I happened to be traveling by train from Moscow to Leningrad. I entered into conversation with a fellow traveler, who turned out to be a professor of political philosophy. Having heard vaguely that the Einstein theory of relativity, which was already often opposed in Germany as "Bolshevism in physics," had been characterized by a number of Soviet scientists as "bourgeois" and "reactionary," and as the strangest rumors circulated about conditions in Russia, I did not know whether this was true. Consequently, I welcomed my chance meeting with this traveling companion and I conversed with him on various problems of political and scientific philosophy, so far as my knowledge of Russian permitted. Finally, half in fun, I said: "I would be grateful if you could answer a question for me. I cannot understand how Einstein's theory of relativity is decried in some countries as 'bolshevistic,' and the same theory is opposed in Russia as 'antibolshevistic.' "

My companion reflected for a moment and then replied categorically: "In the capitalist countries the relativity theory maintains that the capitalistic economic system is only 'relatively right.' On the other hand, in Soviet Russia Einstein's adherents claim that the Communist system is only 'relatively right.' As a result the relativists are quite rightly condemned everywhere."

4. *Einstein's personality and fate symbolize important features of the twentieth century*

Many people have sensed the greatness of Einstein's theories; but men also want to have a reason for the homage that they give. Frequently it is difficult to attain an intellectual understanding of some great achievement, and yet one wants to explain why it is great. As a result, instead of indicating the actual reason for the greatness of the achievement, some superficial, trivial reason is given. Naturally this is not the real reason for the greatness of a man like Einstein.

Despite the superficiality of the reasons given, the widespread admiration for Einstein's theory can be understood only by considering Einstein's position in history. Einstein's theories appeared at an important turning-point in human thought on the universe. People looked with amazement on the revolution occurring before their eyes, and felt that Einstein's theory was a particularly characteristic aspect of this revolution — its nucleus, so to say.

Around 1900 it became continually clearer that even in inanimate matter chemical and physical processes did not occur in accordance with the laws that were valid for a machine, if one understands the word "machine" in the sense in which it is used by the engineer. Explanation of nature in terms of a mechanical analogy, however, had been a characteristic element in the emancipation of the human mind from the bonds of the Middle Ages. In consequence, the collapse of the "mechanical theory of the world" was often interpreted as a failure of this emancipation. It became an argument for a return to the Middle Ages in all aspects of life. More and more people began to doubt whether the progress prophesied by liberalism on the basis of this mechanical science of the nineteenth century would ever be capable of eliminating completely, or even of alleviating substantially, the material and spiritual troubles of mankind. As a result, this feeling of "bankruptcy" in nineteenth-century science went hand in hand with a similar attitude toward liberalism.

Before and after 1900 there appeared characteristic intellectual movements opposed to "materialism" and "liberalism." On the one hand, there was an outcry for a return to the organismic philosophy of the Middle Ages, a trend that in its political form later led to fascism. On the other hand, attempts were made to develop dynamic and organismic forms of materialism, thus giving rise to the various forms of Marxian socialism.

All these philosophical and political groups appealed to the revolution in science for their intellectual basis, and thus in greater or lesser degree to the work and the person of Einstein. Thus, from the standpoint of the history of ideas, he was linked to the roots of fascism and of Marxism. For the same reasons Einstein is linked with those tendencies of the twentieth century which have operated toward a revival of religion, in opposition to the "materialistic" and "atheistic" convictions of eighteenth-century and nineteenth-century science, which were based primarily on the physical science of these periods. Any lessening in the prestige of this science meant a greater prestige for those tendencies which seemed to be directed toward a "new order" in the world.

Einstein is of Jewish descent. At many turning-points in history the Jews have played the part of a scapegoat, so that the fate of this people reveals, as with a magnifying glass, much about the life and the problems of a period. For this ancient people juts into our own time like a rock from some primordial age, and because of its stubborn maintenance of ancient traditions has repeatedly aroused the antagonism of other peoples. With amazement and often with surprise the world sees this ancient stock, from which Jesus had once sprung and which was believed to be dead, producing again and again prophetic types. Karl Marx and Sigmund Freud will perhaps be regarded as false prophets, but no one can deny that they repeatedly agitated and uncovered innermost and sensitive aspects of human nature, thus offending and vexing many men.

In some sense Einstein, too, belongs with these men. H. G. Wells very trenchantly described the characteristic element of his mind as "subtle simplicity." And this "subtle simplicity" has become an essential element of the mind of the twentieth century. The principles underlying Einstein's theories on the physical universe are of an uncanny simplicity, but from them derives a profusion of revolutionary consequences.

For Einstein it was always evident that the history of mankind cannot be reduced to simple formulas. There are so many interacting influences that the future cannot be predicted. In greater or lesser degree we shall always have an impression of chaos, but if we consider the universe as a whole — the sun, the planets, the fixed stars, the Milky Way, and the distant galaxies and stellar nebulæ — there are mathematical laws of such sublime simplicity that the human mind can hardly express its amazement. Einstein has often emphasized that this feeling of amazement is closely related to a feeling of admiration. The existence of these simple laws in the chaos that surrounds us is a mysterious feature of the universe that sometimes arouses in Einstein feelings that he calls "religion." Einstein has expressed the recognition of and the feeling for this "simplicity in the midst of chaos" in the sentence that I have taken as the motto of this book:

"The most incomprehensible thing about the world is that it is comprehensible."

The universe following simple laws and the chaotic world of human action — these are the two interlocking stages on which Einstein's life has been played, and it is this that I want to portray here.

---

5. *Einstein experienced the history of our century*

Einstein passed that period of his life which was significant for the development of his mind in the Germany of Wilhelm II, after the dismissal of Bismarck. His parents belonged to the German lower middle class and shared its admiration for German *Kultur* and the "New Reich." We shall see Einstein already as a small child in vehement protest against the spirit that regarded military drill as a model for education and religious training. We shall see Einstein fleeing the German school to live in a more liberal environment in Switzerland.

The year 1905 was the decisive revolutionary year for the history of the world. In this year the new Russia was developing, Japan was entering the company of the world powers, and the groundwork was being laid for the outbreak of the World War of 1914. In that decisive year of 1905 Einstein still thought little about the germs of the future. At that time, however, he took the decisive steps that were to revolutionize our views of the physical universe. In 1905 he created the theory of relativity, the quantum theory, and the theory of atomic motion. All these revolutions in the picture of the physical world were soon interpreted as revolutions in our philosophical views on matter and energy, space, time, and causality. Soon the profoundest consequences were drawn for the doctrine of free will, for the role of the material and the spiritual in the world, and from this in turn consequences for morality, politics, and religion. Not until forty years later, in 1945, did the technological consequences of Einstein's theories become evident in the liberation of atomic energy.

Later we find Einstein as professor of physics at universities where German was spoken but which were situated outside the German Empire — in Zurich (Switzerland), and in Prague, which then belonged to Austria. In Prague there were already clearly evident the tensions arising from the sympathy of the Slavic population for Russia and France, and the Germanophile orientation of Austrian foreign policy. The atmosphere was pregnant with the coming World War, but Einstein was not much interested then. In this atmosphere he developed his theory of gravitation, and predicted the deviation of stellar light by the attraction of the sun.

In 1913, immediately before the outbreak of the World War, Einstein moved to Berlin. The expectation of participating in the intense scientific life of the city outweighed his aversion to the political and social atmosphere of a Germany led by Prussia, from which he had once fled to Switzerland. We see him in 1914 experiencing the beginning of the World War in Berlin. At first he saw in the war only its pitiable and deplorable human side. He participated in no political movement, even with his sympathies, and worked on his general theory of relativity.

Toward the end of the war, however, especially after the Russian Revolution radically changed the face of the entire struggle, Einstein already felt that he would be driven from his work on the eternal laws of the universe into the world of human chaos. During the war he continued to work on cosmological problems. He investigated the possibility of a hypothesis of the finiteness of the universe, and from these studies developed later the hypothesis of the expanding universe. But the feeling grew stronger within him that the world of chaos was violently overflowing his scientific work. Like many other intellectuals he believed at first that the end of the war would lead to a better world, that cosmic lawfulness would be reflected, at least weakly, in the world of man. In this way the fate of Einstein and his theories became a symbol of the new Europe for which everyone hoped, the Europe of peace and mutual respect among all its peoples and social classes.

Symbolic too, however, was the hatred that his life and work aroused among the enemies of this new Europe. After the war Germany sought scapegoats upon which to satisfy its inhibited aggressive tendencies and to relieve itself of its inferiority complex. These scapegoats were found in the pacifists, Jews, and democrats. As Einstein belonged to all three of these groups and in addition was world-famous, he became a favorite target for attack in Germany, and particularly in Berlin, where he lived. He now saw himself torn out of his world, out of the beautiful dream of an orderly universe ruled by simple laws, of which some slight reflection would appear in the world of men. It was repeatedly thrust upon his consciousness that he belonged to the Jewish people, that he was linked to specific political and social ideals, and that through these circumstances he was involved in the chaotic life of mankind.

We shall read how Einstein courageously assumed this role that was forced on him, assumed it with the same confidence with which he fought his personal struggle for the laws of the universe. He became a leader of the Jewish community and an advocate of international conciliation. He advocated the establishment of a Jewish national home in Palestine, and went on propaganda tours for the Hebrew University in Jerusalem.

We shall see how during the first years after the armistice, from 1919 on, he traveled all over the world to achieve his aims. He visited France, England, Japan, Palestine, and the United States. Everywhere he tried to utilize enthusiasm for the achievements of science in his work for the unity of mankind, and for the goal of finding a worthy place for the Jewish people in the family of nations. These endeavors aroused a wave of hatred against him among the advocates of a "Revolution of the Right." This hatred did not remain confined to the chaotic human world, but overflowed even into the world of theories of the universe. The boundary between the two worlds appeared to become more and more indistinct.

We shall see how Einstein, after returning to Berlin from his travels, found a changed situation; Field Marshal von Hindenburg was now the head of the German Republic. Thenceforth Einstein lived to all intents and purposes in the midst of the brewing National Socialist revolution. The ideas of international conciliation through scientific progress and of liberalism lost prestige, even among scientists.

Stalin's regime in Russia also contributed to this development. Socialism, it was thought, could be erected only on the basis of a powerful, nationalist Russia. This did not fit in with the ideology of the European "liberals."

Einstein now turned away from all party politics, because he found no party actually congenial. He fought from a more ethical, or perhaps even æsthetic, standpoint against militarism, which appeared to him to be the evil spirit of the age. Similarly, he was often unpleasantly affected by the nationalistic spirit of Zionism. He supported Zionism as a movement for the establishment of a Jewish national home, but not as a political party.

Einstein's life enables us to follow very clearly the entire story of the disillusionment of European liberals and intellectuals. We see him turning more and more away from Europe to seek new possibilities for his work across the ocean. Finally we shall see the end of the process that led from the election of Hindenburg as President, by way of the interim governments of Brüning and Papen, to Hitler's accession to power. We shall see how at one stroke the world of human chaos came to rule over the world of knowledge. The authority of the state now declared openly and clearly: the intellect is the servant of the will, science the servant of politics.

In the face of this brutal and naked reality, Einstein himself was forced to give up his radical antimilitaristic position. He was forced to assume a political position toward the struggle in the world and to admit the possibility of a "just" war. We shall see how he becomes an American citizen and experiences the mighty events in this country.

Through Einstein's life one can also trace the intellectual struggles of the twentieth century. As a child he grew up in a milieu that confronted religion with the indifference of the liberal bourgeois of the nineteenth century. In this child, however, religion as a mystical and artistic conception of the world formed an important element of his psychological life, nor did he notice much difference between the Jewish and the Christian religions. Soon after the end of the first World War, however, we find Einstein a conscious member of the Jewish community. Despite his continued aversion to Jewish religious orthodoxy and Jewish nationalism, he saw himself driven to work for this community. He clearly felt the burdens and the psychologically depressed state of the Jews in the world. It was clear to him that these circumstances presaged only greater oppression.

In the period between the two World Wars, Einstein felt and publicly emphasized his solidarity with the Jewish community, but various aspects of his childhood attitude toward religion revived within him. He clearly felt the impotence of science in the face of the increasing influence of the doctrines of the master race and of the glorification of war, and he emphasized that much in the teachings of the Jewish as well as of the Christian churches is opposed to these predominant tendencies. We shall observe the strange spectacle of Einstein being condemned as a materialist, atheist, and bolshevik on the one hand, and on the other being cited enthusiastically and with approval by many Christian and Jewish clergymen as an example of the attitude of the modern scientist to religion. Similarly we shall trace his position in relation to the various political creeds. This attitude was always determined by a purely humanitarian point of view. Like so many progressive people of his time he expected much from the new socialist Russia. We shall see, however, how he was repeatedly repelled by the instances of intolerance that occurred there, and how he remained suspended between hope and fear. The same characteristic is also evident in his relations to the League of Nations, and toward other political ideas, even toward Zionism when it left the realm of ideas and took form in human organizations.

Again and again political and religious parties wanted to interpret Einstein's new physics as splendid confirmation of their philosophical basis and in this way to make Einstein their patron saint. From this we shall acquire a better understanding of the mutual interrelations of science, philosophy, religion, and politics, particularly as they are reflected in the intellectual personality of Einstein, a mirror undimmed and without distortions. In this mind simplicity and subtlety are united in a manner characteristic of great men. We shall learn to understand the manner in which Einstein confronts the world: with absolute confidence in the great things and with inexhaustible distrust of all details.

6. *How dreams come true*

Through the greatest part of his life Einstein's thoughts were in search of the eternal laws ruling the universe while his physical person was subjected to all the disturbances one calls traditionally "world history," by which one means actually the vicissitudes of power politics. This double life reminds us somehow of the medieval church plays, where the Biblical stories were performed on two stages. On the upper stage God the Father, the Holy Virgin, and the angels performed a show of dignity and serenity while on the lower stage human beings displayed the petty miseries of life on earth. This analogy may help us a little to understand Einstein's life during the nationalistic orgy of World War I and the nightmare of the approaching Nazi regime. When Einstein in 1933 had settled peacefully in the United States, it seemed at first that he could now devote himself to his great work of finding the "unified field theory," the system of mathematical laws that would account for all physical facts, electromagnetic as well as mechanical and nuclear phenomena.

After 1939, however, when the active intervention of the United States in the war against the Nazis seemed to be impending, the two stages of the play started merging more and more. Einstein was drawn into the actual performance of "world history." He seemed for many people no longer to be a passive onlooker, but an actor in the united cast of saints and sinners that makes the world's history. The sharp contrasts within his personality pattern, which have baffled so many people and led to so many misjudgments, revealed themselves now on a widely visible stage and made him again a target of futile attacks.

Einstein has fought all his life for peace, friendship, and mutual esteem between all nations, all creeds and races. The establishment of the Nazi government in Germany poisoned the relations between these human groups to a degree that had not been known before among civilized people. A consistent attitude of antiviolence meant in those days an appeal for the suicide of all anti-Nazi forces all over the world. When it became clear that the United States government would actively take sides against the Nazi power, it was obvious that the Nazis would harness the immense potential of German science in the service of their war machine. The extraordinary mind of Franklin D. Roosevelt understood well that the German power could be defeated only by its own weapons.

In the face of the skeptical attitude of a great many politicians and military men, Roosevelt had the vision to realize that the United States had now accumulated enough intellectual strength in science to be a match for the world champion. He gambled on his conviction that seven years of Nazi administration would be enough handicap for German science to ensure a United States victory in the race. There was a project to build "atomic weapons" based upon the most recent theories of physics: the theory of relativity, the quantum theory, and nuclear theory. Einstein, who had made decisive contributions to these theories, expressed the opinion that such weapons are feasible, and Roosevelt followed his advice. The result is known to the world. The first atomic bombs launched upon Japan (1945) put a sudden end to World War II. Forty years before this date (1905) Einstein had announced the law of the conversion of mass into energy and predicted that on this basis in some distant future incredible amounts of energy would be produced. Einstein's theory was regarded then by most "practical people" as a kind of philosophical speculation, which might be very interesting but had no technical application. In 1945 the explosion over Hiroshima exploded also the argument of practical men by which they attempted to minimize the creative imagination of the speculative physicist. Moreover, the philosophy of science proclaimed by the spiritual leaders of the Nazis, about which much will be said in this book, exploded, too, in the mushroom-shaped cloud over Hiroshima. Einstein, the great dreamer among the physicists, and the great scientist among the pacifists, was now actively involved in the making of world history and in the actual political problem of how to avoid an atomic war.

In Einstein's mind, his co-operation with the Zionist movement has been a part of his work for the understanding of nations and creeds. He has thought that the foundation of a home for the homeless Jewish people will appease most of the troubles that their spreading among other peoples has created. In 1948 the immediate goal of the Zionist movement was reached by the foundation of the state of Israel. The Biblical prophets announced again and again that the return of Israel to its old home would be connected with a general peace among the nations, with a general victory of righteousness, and with the end of all quarrels among men.

We read, for instance, in Isaiah lx, 17–18: "I will make thy officers peace, and thine exactors righteousness. Violence shall no more be heard in thy land, wasting nor destruction within thy borders; but thou shalt call thy walls Salvation, and thy gates Praise."

The spirit of these old prophecies was certainly alive in Einstein's mind when he gave his backing to the cause of Zionism. But when in 1948 these dreams became true, it happened in an atmosphere of war and of hatred between the nations in the Holy Land. The rebirth of Israel as well as the conquest of atomic energy had been envisaged by the intellectual and spiritual leaders of mankind as milestones on the way to perpetual peace and happiness; but both materialized as a result of the struggle for power among the nations and have been surrounded from the start by clouds of mutual suspicion.

A great many people separate themselves from noble causes because their actual realization is tarnished by a good deal of meanness and foolishness. Einstein never made use of such pretexts, which are often hypocritical excuses for a lack of courage. Einstein has always been deeply convinced that there are pure results only in the realm of creative imagination. He has never been intimidated into denouncing the living results of his imagination because these results were affected by the lawlessness of all actual happening in the realm of human affairs. Every honest fighter for good causes has to be aware of this situation, which the great fighter for peace, according to the gospels, characterized by the words: "Render unto Cæsar the things which are Cæsar's, and unto God the things that are God's."

Einstein

His Life and Times

I

EINSTEIN'S YOUTH AND TRAINING

---

1. *Family Background*

As far back as Einstein's memory extends, both his paternal and his maternal ancestors lived in small towns and villages of Swabia in southwestern Germany. They were small merchants, shopkeepers, and artisans, and none of them had ever attracted attention in consequence of any intellectual achievement. Einstein himself remarks on occasions when he is questioned about his ancestors: "The circumstances under which they lived were much too restricted to permit them to distinguish themselves." They did not stand out in very bold relief against their environment.

This background of southwestern Germany is very important in the understanding of Einstein's character. The Swabians merge almost imperceptibly with the French through the mediation of the neighboring Alsatians, and they are reflective, are practical in daily life, and participate joyfully in every kind of art and pleasure as well as in philosophical and religious speculations; but they are averse to any kind of mechanical order. Their nature is different from that of the sober, practical Prussians, interested in order and domination, and from that of the earthy, merry, sometimes rather coarse Bavarians.

The differences in the character of these people are quite evident from their dialects. The Swabian speech is melodic and flows slowly like a rippling, murmuring brook, unlike that of the Germans of the ruling class of officers and officials, which sounds like a bugle in a military camp. Neither is it like the cynical bleating of the Berliner, critical of everything in heaven or on earth, nor like the pompously precise literary German of the pastors and professors.

Elements of this friendly dialect are still to be heard in Einstein's speech, although it has now been greatly blurred and almost obliterated by his travels in many different countries. In particular his speech contains an admixture of certain tones derived from the Swiss, which is indeed related to the dialect of

3

southwestern Germany, but has a somewhat rougher tone. But anyone listening to the speech of Einstein's second wife, who is descended from the same family as her husband, could still hear the genuine, pleasantly agreeable Swabian idiom. For her he was always "Albertle," the land was *"ländle,"* the city (*Stadt*) *"Städtle."* Everything received the diminutive suffix *"le,"* which gives the dialect a quality of tenderness and affection.

The fact that Einstein's ancestors were Jewish made a difference, but not to so large an extent as one might expect. During the period when his parents were growing up, the Jews in these small towns of Swabia did not differ greatly from the rest of the population in their mode and way of life. They no longer clung so firmly to their complicated customs and usages, which rendered difficult the growth of any intimacy between them and the rest of the population; and with the disintegration of these barriers they tended, to an ever increasing degree, to lose their position as a separate and unique group. The life of the Jews in these districts was not similar to that in Berlin, where there was a class of rich, educated Jews, who themselves developed a specific variant of Berlin culture. There was none of this in the small Swabian towns. Here the Jews, like the other inhabitants, led a quiet life, associated with their natural environment, and were but little influenced by the nervous hustle and bustle of the metropolis.

In progressive circles at this time the reading of the Bible and other books dealing with the doctrines of the Jews was no longer the only source of truth. The Bible was read like other belles-lettres and edifying literature, and in Jewish families the classical German authors appeared beside the prophets as teachers of morality and conduct. Schiller, Lessing, and Heine were honored like the preacher Solomon and the Book of Job. Friedrich Schiller particularly, with his moral, almost Biblical pathos and glorification of a general love of mankind, became extremely popular among Jewish families and was an important element in the education of their children. That he was a Swabian was another reason for regarding him as something of a close relative. In Einstein's family this Schiller cult and the admiration for the Enlightenment, which was bound up with it, also played a large part in the training of the younger generation.

The writer Berthold Auerbach, who was active between 1840 and 1870, is perhaps characteristic of the life and intellectual temper of the Jews in Swabia in the time of Einstein's parents and grandparents. He was the first to portray the daily life of the peasants of the Black Forest. These *Schwarzwälder Dorfgeschichten* (*Tales of the Black Forest*) are probably somewhat too idealized and artificial for present-day taste, but they were considered by contemporaries to be a gratifying counterpoise to what was later called "Berlin gutter literature" and was regarded as a characteristic contribution of the Jews to German literature.

It must also be mentioned that after 1871, as a result of the Franco-Prussian War, Prussia became the predominant power in Germany, and has profoundly affected and influenced the character of the Germans. The unification of the majority of the German tribes and the restoration of a powerful German Empire was not initiated by the intellectual class. Writers and scholars had long dreamed and sung of this goal, but they had hoped that, as the Swabian poet Uhland put it, "the Imperial crown of the new Germany would be anointed with a drop of democratic oil." But the dream did not come true. Bismarck carried out his work not with "democratic oil," but with "blood and iron," and with methods that were opposed by all the intellectually progressive groups in Germany. Furthermore, the new Germany did not arise from the national elements possessing an older culture: the Swabians, Rhinelanders, and Austrians who had produced Schiller, Goethe, Mozart, and Beethoven. The rulers came from the tribes of the east, which were composed of those who had settled on soil won by conquering, Germanizing, and partially exterminating the original Slavic population, and of those descended from the subjugated tribes. They thus formed an amalgamation of oppressors and oppressed well able to command and obey.

This situation placed the intellectual groups of all Germany, and particularly those of the older, cultural sections, in an ambiguous and partially mortifying position. They could not avoid admitting that the methods of the new rulers were more effective than their own, since they had been so successful; yet they could not overcome their aversion to the adoration of force and the glorification of order as ends in themselves. Such an empty mechanical arrangement of life was repugnant to them, with their inclination and aptitude for art and science. The new masters were not to their liking, but they were compelled to admire and to some extent to imitate them. The German scholars acquired a feeling of inferiority toward the Prussian officers, and learned to restrict themselves to their own "subject," to leave

public life to the ruling group as their "subject," and to stand
at attention, even intellectually, at the sound of a commanding voice.

All this was equally true of the Jews. They, too, admired the
new Empire and the energetic methods of its rulers. Even
though in their homes they cultivated the intellectual tradition
of the Jews and of the German classical period, yet in public
life they tried to assimilate themselves to the ruling class in conduct and ideas.

Only for those who were strong enough not to accord recognition to outward success and who could not be compensated by
any external manifestation of power for the loss of freedom and
the cultural atmosphere was it possible to maintain an independent attitude and to resist the prevailing trend. We shall see
that from his youth Einstein belonged to these people. Even
though later he frequently came into conflict with the prevailing tendencies in Germany, yet he always retained a certain
attachment for his Swabian homeland and its people.

2. *Childhood*

Albert Einstein was born on March 14, 1879 at Ulm,
a middle-sized city in the Swabian part of Bavaria. This city is
of no significance in his life, however, since a year after his birth
the family moved to Munich. A year later a daughter was born
and there were no other children. Munich, the city in which
Albert spent his youth, was the political and intellectual center
of southern Germany. Thus the family had already departed
from the romantic nooks of Swabia and had made a transition
to a more urban life. Their house, however, was a cottage surrounded by a large garden in the suburbs. Albert's father, Hermann Einstein, had a small electrochemical factory that he
operated with the aid of his brother, who lived with the family.
The former attended to the commercial side of the enterprise,
while the latter acted as technical director.

Hermann Einstein was an optimistic person who enjoyed life.
He was not a particularly good business man and was frequently
unsuccessful, but such failures did not change his general outlook on life. His mode of life and his *Weltanschauung* differed
in no respect from those of the average citizen in that locality.
When his work was done, he liked to go on outings with his

6

family into the beautiful country around Munich, to the romantic lakes and mountains, and he was fond of stopping at the pleasant, comfortable Bavarian taverns, with their good beer, radishes, and sausages. Of the traditional Jewish fondness for reading edifying literature he had retained only a love for German poetry, especially that of Schiller and Heine. The dietary laws and other customary usages of the Jewish community were to him only an ancient superstition, and in his house there was no trace of any Jewish custom. Or, to put it more concisely, the ancient customs themselves had disappeared, but several humane usages connected with them were retained. For instance, every Thursday the Einstein family invited a poor Jewish student from Russia to share their midday meal with them — a reflection, no doubt, of the old Sabbath custom. Similarly, their preference for the dramas and poems of Schiller, replete with moral pathos, was a substitute for the reading of the Bible. In his political views, too, Einstein's father, like most others, was afraid of the dominant Prussians, but admired the new German Empire, its Chancellor Bismarck, General Moltke, and the old Emperor Wilhelm I.

Einstein's mother, born Pauline Koch, was of a more serious and artistic nature, with a fine sense of humor. But the rather meager material conditions under which she lived led her to be satisfied with a tolerably secure existence for herself and her children. She found much happiness and consolation in her music, and when engineers from the factory dropped in for an evening visit, they accompanied her on the piano. Above all she loved German classical music, especially Beethoven's piano sonatas.

The uncle who lived with the family was a man whose interest in the more refined aspects of intellectual life was greater than that of the father. He was a trained engineer, and it was from him that Albert received his first impulsions in mathematics.

There can be no doubt that this origin in a provincial, semi-rural milieu was of the greatest significance for Albert Einstein's entire psychological development. He has never become a completely urban person. He was always somewhat afraid of Berlin and later also of New York. Connected with this attitude is a certain trait that characterizes his artistic taste and that certainly appeared old-fashioned to modern Berliners. Einstein's preference for the German classics in literature as well as in music was expressed at a time when the intellectual circles of Berlin declared that such tastes had long since been superseded. His predilection for Schiller is a particularly characteristic feature, by which one recognizes a member of a culture not that of twentieth-century Berlin.

On the whole, little Albert was no child prodigy. Indeed, it was a very long time before he learned to speak, and his parents began to be afraid that he was abnormal. Finally the child did begin to speak, but he was always taciturn and never inclined to enter into the games that nursemaids play with children in order to keep both the children and themselves in good humor. A governess entrusted with Albert's childhood training even dubbed him *Pater Langweil* (Father Bore). He did not like any strenuous physical exertions such as running and jumping, perhaps for the reason that he considered himself too weak for such activities. From the very beginning he was inclined to separate himself from children of his own age and to engage in daydreaming and meditative musing.

He disliked particularly playing at being a soldier, which the children of most countries engage in with the greatest delight, and which especially in the Germany of Bismarck and Moltke was imbued with an almost mythical splendor. When the soldiers marched through the streets of Munich accompanied by the roll of drums and the shrill of fifes, a combination, characteristic of the German army, that gives the music an exciting, compelling rhythm and a wild tonal quality, and when the pavements and the windowpanes rattled from the pawing of the horses' hoofs, the children enthusiastically joined the parade and tried to keep in step with the soldiers. But when little Albert, accompanied by his parents, passed such a parade, he began to cry. In Munich parents would often tell their children: "Some day, when you grow big, you, too, can march in the parade," and most boys were spurred to greater and more ambitious efforts by this prospect. Albert, however, said to his parents: "When I grow up I don't want to be one of those poor people." When the majority saw the rhythm of a happy movement, he observed the coercion imposed upon the soldiers; he saw the parade as a movement of people compelled to be machines.

At this time Einstein apparently already revealed one of his most characteristic traits: his intractable hatred of any form of coercion arbitrarily imposed by one group of people on another. He detested the idea of the oppressor preventing the oppressed from following their inclinations and developing their natural talents, and turning them into automatons. On the other hand Albert was also conscious of the natural laws of the universe; he felt that there are great eternal laws of nature. As a child he **was** able to understand them only in the form of traditional religion, and felt attracted toward it and its ritual precepts, which symbolized a feeling for the laws of the universe. He was offended by the fact that his father always scoffed at religion, and he regarded this derision as resulting from a type of thought that is in a certain sense disharmonious and refuses to submit to the eternal laws of nature. This dual attitude — hatred for the arbitrary laws of man and devotion to the laws of nature — has accompanied Einstein throughout his life and explains many of his actions that have been considered peculiar and inconsistent.

At that time the German elementary schools were conducted on a denominational basis, the clergy of each religious group controlling its schools. Since Munich was for the most part Catholic, most of the schools were naturally of that denomination. Nominally Einstein's parents probably adhered to the Jewish religion, but they were not sufficiently interested in a Jewish education to send their children to a Jewish school since there was none near their home and it would have been expensive. His parents may even have felt that by sending their boy to a Catholic school he would come into more intimate contact with non-Jewish children. At any rate, Albert attended the Catholic elementary school, where he was the only Jew in his class.

Young Albert experienced no unpleasantness because of this situation. There was only a slight feeling of strangeness resulting naturally from the different religious traditions, and this factor was definitely of secondary significance and did not increase to any marked degree his difficulty in forming intimate friendships with his fellow pupils. The difficulty was due fundamentally to his character.

Albert received regular instruction in the Catholic religion and he derived a great deal of pleasure from it. He learned this subject so well that he was able to help his Catholic classmates when they could not answer the teacher's questions immediately. Einstein has no recollection of any objection having arisen to the participation of a Jewish pupil in Catholic religious instruction. On one occasion the teacher attempted a somewhat strange kind of object lesson by bringing a large nail to the class and telling the pupils: "The nails with which Christ was nailed to the cross looked like this." But he did not add, as sometimes happens, that the Crucifixion was the work of the Jews. Nor did the idea enter the minds of the students that because of this they must change their relations with their classmate Albert. Nevertheless Einstein

9

found this kind of teaching rather uncongenial, but only because it recalled the brutal act connected with it and because he sensed correctly that the vivid portrayal of brutality does not usually intensify any sentiments of antagonism to it but rather awakens latent sadistic tendencies.

It was very characteristic of young Einstein's religious feeling that he saw no noticeable difference between what he learned of the Catholic religion at school and the rather vaguely remembered remnants of the Jewish tradition with which he was familiar at home. These elements merged in him into a sense of the existence of lawfulness in the universe and in the representation of this harmony by means of different kinds of symbols, which he judged rather on the basis of their æsthetic value than as symbols of the "truth."

On the whole, however, Einstein felt that school was not very different from his conception of barracks — that is, a place where one was subject to the power of an organization that exercised a mechanical pressure on the individual, leaving no area open within which he might carry on some activity suited to his nature. The students were required to learn mechanically the material presented to them, and the main emphasis was placed on the inculcation of obedience and discipline. The pupils were required to stand at attention when addressed by the teacher and were not supposed to speak unless asked a question. Independent questions addressed by students to the teacher and informal conversations between them were rare.

Even when Albert was nine years old and in the highest grade of the elementary school, he still lacked fluency of speech, and everything he said was expressed only after thorough consideration and reflection. Because of his conscientiousness in not making any false statements or telling lies he was called *Biedermeier* (Honest John) by his classmates. He was regarded as an amiable dreamer. As yet no evidence of any special talent could be discovered, and his mother remarked occasionally: "Maybe he will become a great professor some day." But perhaps she meant only that he might develop into some sort of eccentric.

3. *Gymnasium in Munich*

At the age of ten, young Einstein left the elementary school and entered the Luitpold Gymnasium in Munich. In Germany the period between the ages of ten and eighteen, the years that are of decisive importance in the intellectual development of adolescents, are spent in the gymnasium. The aim of these institutions was to give the young people a general education based upon the acquisition of ancient Greek and Roman culture, and for this purpose most of the time was devoted to learning Latin and Greek grammar. Because of the complications of these subjects, and since the students were required to learn all the rules pedantically, little time was left to acquire a real understanding of the culture of antiquity. Furthermore, it would have been a much more difficult task for the majority of the teachers. It was claimed that the process of learning the grammar of one or two complicated languages is an indispensable training for the mind and a disciplining of the intellect hardly attainable otherwise. For Einstein, however, aspiring to learn the laws of the universe, this mechanical learning of languages was particularly irksome, and this kind of education seemed very much akin to the methods of the Prussian army, where a mechanical discipline was achieved by repeated execution of meaningless orders.

Later, when speaking about his impressions of school, Einstein frequently said: "The teachers in the elementary school appeared to me like sergeants, and the gymnasium teachers like lieutenants." The sergeants in the German army of Wilhelm II were notorious for their coarse and often brutal behavior toward the common soldiers, and it was well known that, with the troops completely at their mercy, sadistic instincts developed in them. The lieutenants, on the other hand, being members of the upper class, did not come into direct contact with the men, but they exerted their desire for power in an indirect manner. Thus when Einstein compared his teachers to sergeants and lieutenants, he regarded their tasks to be the inculcation of a certain body of knowledge and the enforcing of mechanical order upon the students. The pupils did not view the teachers as older, more experienced friends who could be of assistance to them in dealing with various problems of life, but rather as superiors whom they feared and tried to predispose favorably to themselves by behaving as submissively as possible.

There was one teacher in the gymnasium, named Ruess, who really tried to introduce the students to the spirit of ancient culture. He also showed them the influence of these ancient ideas in the classical German poets and in modern German culture. Einstein, with his strong feeling for everything artistic and for all ideas that brought him closer to the hidden harmony of the world, could hardly have enough of this teacher. He aroused in him a strong interest in the German classical writers, Schiller and Goethe, as weil as in Shakspere. The periods devoted to the reading and discussion of *Hermann und Dorothea,* Goethe's half-romantic, half-sentimental love story written in a period of the greatest political unrest, remained deeply engraved in Einstein's memory. In the gymnasium the students who had not completed their assignments were punished by being made to stay after school under the supervision of one of the teachers. In view of the tedious and boring character of the ordinary instruction, these extra periods were regarded as a real torture. But when Ruess conducted the extra period, Einstein was happy to be punished.

The fact that in the midst of all the mechanical drilling he was sometimes able to spend an hour in an artistic atmosphere made a great impression on him. The recollection of this class remained very vivid in his mind, but he never stopped to consider what sort of impression he had made on the teacher. Many years later, when he was already a young professor at Zurich, Einstein passed through Munich and, overcome by his memories of the only man who had really been a teacher to him, decided to pay him a visit. It seemed obvious to him that the teacher would be happy to learn that one of his students had become a professor. But when Einstein arrived at Ruess's quarters dressed in the careless manner that was characteristic of him then as well as later, Ruess had no recollection of any student named Einstein and could not comprehend what the poorly dressed young man wanted of him. The teacher could only imagine that by claiming to be one of his former pupils the young man thought he could borrow money from him. Apparently it never entered Ruess's mind that a student could pay him a visit to express a feeling of gratitude for his teaching. It is possible that his teaching had not been so good as it appeared in Einstein's memory and perhaps he had only imagined it. But in any case the visit was very embarrassing for Einstein and he departed as quickly as possible.

4. *Intellectual Interests*

When Einstein was five years old his father showed him a pocket compass. The mysterious property of the iron needle that always pointed in the same direction no matter how the compass case was turned made a very great impression on the young child. Although there was nothing visible to make the needle move, he concluded that something that attracts and turns bodies in a particular direction must exist in space that is considered empty. This was one of the impressions which later led Einstein to reflect on the mysterious properties of empty space.

As he grew up, his interest in natural science was further aroused by the reading of popular scientific books. A Russian Jewish student who ate at Einstein's home on Thursdays called his attention to Aaron Bernstein's *Naturwissenschaftliche Volks-bücher* (*Popular Books on Natural Science*), which were widely read by laymen interested in science about that time. These books discussed animals, plants, their mutual interdependence, and the hypotheses concerning their origin; they dealt with stars, meteors, volcanoes, earthquakes, climate, and many other topics, never leaving out of sight the greater interrelation of nature. Soon Einstein was also an enthusiastic reader of such books as Büchner's *Kraft und Stoff* (*Force and Matter*), which attempted to gather together the scientific knowledge of the time and to organize it into a sort of complete philosophical conception of the universe. The advocates of this view, frequently called "materialism" although it should rather be called "naturalism," wanted to understand and explain all celestial and terrestrial occurrences by analogy with the natural sciences and were particularly opposed to any religious conception of the nature of the universe.

Today such books as Büchner's *Kraft und Stoff* are considered superficial and we may wonder how at that time young people like Einstein who were capable of independent thought could have been stirred by them. Yet if we have any sense of historical values and justice, we should ask ourselves what recent books are to be regarded as the analogues of those earlier works. In reply we can point to such books as Sir James Jeans's *The Mysterious Universe*. Probably a really critical judge would not be able to say that Büchner's book is more superficial than those of similar contemporary writers. At any rate, we find a very good popular presentation of the scientific results themselves, and a rather vague philosophical interpretation, which may be accepted or not according to one's taste.

Einstein's interest in mathematics was also aroused at home and not at school. It was his uncle and not the teacher at the gymnasium who gave him his first understanding of algebra.

"It is a merry science," he told the boy; "when the animal that we are hunting cannot be caught, we call it *x* temporarily and continue to hunt it until it is bagged." With such instruction, Albert found a great deal of pleasure in solving simple problems by hitting upon new ideas instead of just using a prescribed method.

He was impressed most, however, when at about the age of twelve he obtained for the first time a systematic textbook on geometry. It was a book to be used in a class that had just started, and, like many children who are curious about the new subjects they are going to take up in school, he tried to delve into the subject before it acquired the unpleasant and irksome quality that teachers generally imparted. Having begun to read the book, he was unable to put it down. The clarity of the exposition and the proof given for every statement, as well as the close connection between the diagrams and the reasoning, impressed him with a kind of orderliness and straightforwardness that he had not encountered before. The world with its disorder and uncleanliness suddenly appeared to him to contain also an element of intellectual and psychological order and beauty.

Ever since Albert was six years old his parents had insisted that he take violin lessons. At first this was only another kind of compulsion added to the coercion of the school, as he had the misfortune to be taught by teachers for whom playing was nothing but a technical routine, and he was unable to enjoy it. But when he was about thirteen years old he became acquainted with Mozart's sonatas and fell in love with their unique gracefulness. He recognized that his technique was not equal to the performance of these compositions in the light-handed manner necessary to bring out their essential beauty, and he attempted repeatedly to express their light, carefree grace in his playing. In this way, as a result of his efforts to express a particular emotional mood as clearly as possible and not through technical exercises, he acquired a certain skill in playing the violin and a love for music, which he has retained throughout his life. The feeling of profound emotion that he experienced in reading the geometry books is perhaps to be compared only with his experience as a fourteen-year-old boy when for the first time he was able to take an active part in a chamber-music performance.

At the age of fourteen, while he was still reading Büchner's books, Einstein's attitude toward religion experienced an important change. While in the elementary school he had received Catholic instruction, in the gymnasium he received instruction in the Jewish religion, which was provided for the students of this sect. Young Einstein was greatly stirred by the comments of the teachers of religion on the Proverbs of Solomon and the other parts of the Old Testament dealing with ethics. This experience made a permanent impression and left him with a profound conviction of the great ethical value of the Biblical tradition. On the other hand Einstein saw how the students were compelled to attend religious services in Jewish temples whether they had any interest in them or not. He felt that this did not differ from the coercion by means of which soldiers were driven to drill on the parade ground, or students to unravel subtly invented grammatical puzzles. He was no longer able to regard ritual customs as poetic symbols of the position of man in the universe; instead he saw in them, more and more, superstitious usages preventing man from thinking independently. There arose in Einstein an aversion to the orthodox practices of the Jewish or any other traditional religion, as well as to attendance at religious services, and this he has never lost. He made up his mind that after graduation from the gymnasium he would abandon the Jewish religious community and not become a member of any other religious group, because he wanted to avoid having his personal relationship to the laws of nature arranged according to some sort of mechanical order.

5. *Departure from Munich*

When Einstein was fifteen an event occurred that diverted his life into a new path. His father became involved in business difficulties, as a result of which it appeared advisable to liquidate his factory in Munich and seek his fortune elsewhere. His pleasure-loving, optimistic temperament led him to migrate to a happier country, to Milan in Italy, where he established a similar enterprise. He wanted Albert, however, to complete his studies at the gymnasium. At this time it was axiomatic for every middle-class German that an educated person must have a diploma from a gymnasium, since only this diploma entitled him to become a student at a university. And as a course of study leading to a degree was in turn necessary before one could obtain a position in one of the intellectual professions, Einstein, like all the others, felt compelled to complete his course at the gymnasium.

In the field of mathematics Einstein was far ahead of his fellow students, but by no means so in classical languages. He felt miserable at having to occupy himself with things in which he was not interested but which he was supposed to learn only because he had to take an examination in them. This feeling of dissatisfaction grew greater when his parents departed and left him in a boarding-house. He felt himself a stranger among his fellow students and regarded their insistence upon his participation in all forms of athletic activities as inconsiderate and coarse. He was probably friendly to all, but his skeptical attitude toward the organization and the spirit of the school as a whole was quite clear to the teachers and students and aroused a sense of uneasiness in many of them.

As he developed into an independently thinking man, the thought of having to submit for some time yet to the pedagogical methods of the gymnasium became more and more unbearable. Although he was good-natured and modest in personal intercourse, nevertheless then as well as later he stubbornly defended his intellectual life against the entry of any external constraint. He found it more and more intolerable to be compelled to memorize rules mechanically, and he even preferred to suffer punishment rather than to repeat something he had learned by rote without understanding.

After half a year of suffering in solitude Einstein tried to leave the school and follow his parents to Italy. To Einstein, living in Munich, which was dominated by the cold, rigid Prussian spirit, colorful Italy, with its art- and music-loving people living a more natural and less mechanized life, appeared to be a beckoning paradise. He worked out a plan that would enable him to run away from school, at least for a while, without forfeiting his chances of continuing his studies. Since his knowledge of mathematics was far ahead of the requirements of the gymnasium, he hoped that he might perhaps be admitted to a foreign institute of technology even without a diploma. He may even have thought that, once he was out of Germany, everything would take care of itself.

From a physician he obtained a certificate stating that because of a nervous breakdown it was necessary for him to leave school for six months to stay with his parents in Italy, where he could recuperate. He also obtained a statement from his mathematics teacher affirming that his extraordinary knowledge of mathematics qualified him for admission to an advanced institution for the study of such subjects. His departure from the gymnasium was ultimately much easier than he had anticipated. One day his teacher summoned him and told him that it would be desirable if he were to leave the school. Astonished at the turn of events, young Einstein asked what offense he was guilty of. The teacher replied: "Your presence in the class destroys the respect of the students." Evidently Einstein's inner aversion to the constant drill had somehow manifested itself in his behavior toward his teachers and fellow students.

On arriving at Milan he told his father that he wanted to renounce his German citizenship. His father, however, kept his own, so that the situation was rather unusual. Also, since Einstein could not acquire any other citizenship immediately, he became stateless. Simultaneously he renounced his legal adherence to the Jewish religious community.

The first period of his stay in Italy was an ecstasy of joy. He was enraptured by the works of art in the churches and in the art galleries, and he listened to the music that resounded in every corner of this country, and to the melodic voices of its inhabitants. He hiked through the Apennines to Genoa. He observed with delight the natural grace of the people, who performed the most ordinary acts and said the simplest things with a taste and delicacy that to young Einstein appeared in marked contrast to the prevalent demeanor in Germany. There he had seen human beings turned into spiritually broken but mechanically obedient automatons with all the naturalness driven out of them; here he found people whose behavior was not so much determined by artificial, externally imposed rules, but was rather in consonance with their natural impulses. To him their actions appeared more in accord with the laws of nature than with those of any human authority.

This paradisal state of delight, however, could exist only as long as Einstein was able to forget completely — as he did for a while — the urgent demands that the practical necessities of life made upon him. The need for a practical occupation was particularly urgent since his father was again unsuccessful in Italy. Neither in Milan nor in Pavia did his electrical shop succeed. Despite his optimism and happy outlook on life, he was compelled to tell Albert: "I can no longer support you. You will have to take up some profession as soon as possible." The pressure that had hardly been released appeared to have returned. Had his departure from the gymnasium been a disastrous step? How could he return to the regular path leading to a profession?

Einstein's childhood experience with the magnetic compass

had aroused his curiosity in the mysterious laws of nature, and his experience with the geometry book had developed in him a passionate love for everything that is comprehensible in terms of mathematics and a feeling that there was an element in the world that was completely comprehensible to human beings. Theoretical physics was the field that attracted him and to which he wanted to devote his life. He wanted to study this subject because it deals with the question: how can immeasurably complicated occurrences observed in nature be reduced to simple mathematical formulæ?

With his interest in the pure sciences of physics and mathematics and the training required for a more practical profession, together with the fact that his father was engaged in a technical occupation, it seemed best that young Einstein should study the technological sciences. Furthermore, since he lacked a diploma from a gymnasium but had an excellent knowledge of mathematics, he believed that he could more easily obtain admission to a technical institution than to a regular university.

---

6. *Student at Zurich*

At that time the most famous technical school in central Europe outside of Germany was the Swiss Federal Polytechnic School in Zurich. Einstein went there and took the entrance examination. He showed that his knowledge of mathematics was far ahead of that of most of the other candidates, but his knowledge of modern languages and the descriptive natural sciences (zoology and botany) was inadequate, and he was not admitted. Now the blow had fallen. What he had feared ever since leaving Munich had come to pass and it looked as though he would be unable to continue in the direction he had planned.

The director of the Polytechnic, however, had been impressed by Einstein's knowledge of mathematics and advised him to obtain the required diploma in a Swiss school, the excellent, progressively conducted cantonal school in the small city of Aarau. This prospect did not appeal very much to Einstein, who feared that he would again become an inmate of a regimented institution like the gymnasium in Munich.

Einstein went to Aarau with considerable misgiving and apprehension, but he was pleasantly surprised. The cantonal school was conducted in a very different spirit from that of the Munich

Abraham Rupert Einstein and Helen Einstein, grandparents of the scientist

Einstein as a child

Einstein and his sister Maja

gymnasium. There was no militaristic drilling, and the teaching was aimed at training the students to think and work independently. The teachers were always available to the students for friendly discussions or counsel. The students were not required to remain in the same room all the time, and there were separate rooms containing instruments, specimens, and accessories for every subject. For physics and chemistry there were apparatuses with which the student could experiment. For zoology there were a small museum and microscopes for observing minute organisms, and for geography there were maps and pictures of foreign countries.

Here Einstein lost his aversion to school. He became more friendly with his fellow students. In Aarau he lived with a teacher of the school who had a son and a daughter with whom Einstein made trips to the mountains. He also had an opportunity to discuss problems of public life in detail with people who, in accordance with the Swiss tradition, were greatly interested in such affairs. He became acquainted with a point of view different from that which he had been accustomed to in Germany.

After one year at the cantonal school Einstein obtained his diploma and was thereupon admitted to the Polytechnic School in Zurich without further examination. In the meantime, however, he had abandoned the plan of taking up a practical profession. His stay at Aarau had shown him that a position as a teacher of physics and mathematics at an advanced school would permit him to pursue his favorite studies and at the same time enable him to make a modest living. The Polytechnic had a department for training teachers in physical and mathematical subjects, and Einstein now turned to this pursuit.

During the year at the cantonal school Einstein had become certain that the actual object of his interest was physics and not pure mathematics as he had sometimes believed while still in Munich. His aim was to discover the simplest rules by which to comprehend natural laws. Unfortunately, at that time it was just this teaching of physics that was rather outdated and pedantic at the Polytechnic. The students were merely taught the physical principles that had stood the test of technical applications and been accepted in all the textbooks. There was little if any objective approach to natural phenomena, or logical discussion of the simple comprehensive principles underlying them.

Even though the lectures on physics were not marked by any profundity of thought, they did stimulate Einstein to read the works of the great investigators in this field. Just about this time, at the end of the nineteenth century, the development of physical science had reached a turning-point. The theories of this period had been written in stimulating form by the outstanding scientists. Einstein devoured these classics of theoretical physics, the works of Helmholtz, Kirchoff, Boltzmann, Maxwell, and Hertz. Day and night Einstein buried himself in these books, from which he learned the art of erecting a mathematical framework on which to build up the structure of physics.

The teaching of mathematics was on a much higher level. Among the instructors was Hermann Minkowski, a Russian by birth, who, although still a young man, was already regarded as one of the most original mathematicians of his time. He was not a very good lecturer, however, and Einstein was not much interested in his classes. It was just at this time that Einstein lost interest in pure mathematics. He believed that the most primitive mathematical principles would be adequate to formulate the fundamental laws of physics, the task that he had set for himself. Not until later did it become clear to him that the very opposite was the case: that for a mathematical formulation of his idea concepts derived from a very highly developed type of mathematics were required. And it was Minkowski, whose mathematical lectures Einstein found so uninteresting, who put forth ideas for a mathematical formulation of Einstein's theories that provided the germ for all future developments in the field.

At this period the Polytechnic enjoyed a great international reputation and had a large number of students from foreign countries. Among them were many from eastern and southeastern Europe who could not or would not study in their native countries for political reasons, and hence Zurich became a place where future revolutions were nurtured. One of these with whom Einstein became acquainted was Friedrich Adler, from Austria. He was a thin, pale, blond young man who like other students from the east united within himself an intense devotion to his studies and a fanatical faith in the revolutionary development of society. He was the son of Viktor Adler, a leading Social Democrat politician of Vienna, who tried to keep his son out of politics by sending him to study physics at Zurich.

Another of Einstein's acquaintances was Mileva Maritsch, a young woman from Hungary. Her mother tongue, however, was Serbian and she professed the Greek Orthodox religion. She belonged to that group of her people who lived in considerable numbers in southeastern Hungary and always carried on a violent struggle against the Magyar domination. Like many of the women students from eastern Europe, she paid attention only to her work and had few opportunities to attract the attention of men. She and Einstein found a common interest in their passion for the study of the great physicists, and they spent a great deal of time together. For Einstein it had always been pleasant to think in society, or, better perhaps, to become aware of his thoughts by putting them into words. Even though Mileva Maritsch was extremely taciturn and rather unresponsive, Einstein in his zeal for his studies hardly noticed this.

This student period at Zurich, which was so important for Einstein's mental development, was not such an easy time for him with regard to practical living. His father's financial situation was so difficult that he could not contribute anything to his son's support. Einstein received one hundred Swiss francs monthly from a wealthy relative, but he had to put aside twenty of these every month to accumulate the fee necessary for the acquisition of Swiss citizenship, which he hoped to obtain soon after graduation. He did not experience any real material hardships, but on the other hand he could not afford any luxuries.

7. *Official of a Patent Office*

Einstein completed his studies just at the turn of the century and now faced the necessity of seeking a position. When a young man with extraordinary interest and ability in science has completed the regular course of study at a university or technical academy, it is important and generally desirable for him to obtain further training to become an independent investigator by acting as assistant to a professor at a university. In this way he learns the methods both of teaching and of making scientific investigations by working under an experienced person. Since this appeared to be the appropriate path for him, Einstein applied for such a position. It became evident, however, that the same professors who had praised his scientific interest and talent so highly had no intention whatever of taking him on as an assistant. Nor did he receive any direct explanation of this refusal.

With no possibility of a teaching position at the Polytechnic, the only alternative was to look for one in a secondary school. Here again, despite excellent letters of recommendation from his professors, he was unsuccessful. The only thing he obtained was a temporary position in a technical vocational school at Winterthur, and after a few months he was again unemployed.

It was now 1901. Einstein was twenty-one and had become a Swiss citizen. Through a newspaper he found that a gymnasium teacher in Schaffhausen, who maintained a boarding-house for students, was looking for a tutor for two boys. Einstein applied for the job and was hired. Thus he came to this small city on the Rhine whose famous waterfalls resounded throughout the vicinity and where numerous tourists stopped to see the natural phenomenon, which received three stars in the Baedeker.

Einstein was not dissatisfied with his work. He enjoyed molding the minds of young people and trying to find better pedagogical methods than those he had been accustomed to in school. But he soon noticed that other teachers spoiled the good seed he sowed, and he asked that the teaching of the two boys be left completely in his hands. One can well imagine that the gymnasium teacher who conducted the boarding-house regarded this request as a rebellion against his authority. He felt there was an atmosphere of revolt and discharged Einstein. By this action Einstein now realized that it was not only the students but teachers as well who were crushed and made pliable by the mechanical treadmill of the ordinary school.

Einstein was again in a difficulty. All his efforts to find a teaching position failed despite the fact that he held a diploma from the Polytechnic and Swiss citizenship papers. He himself could not quite understand the reason for his failure. It may have been that he was not regarded as a genuine Swiss. With his recent citizenship, he was what the genuine Swiss patriots called a "paper Swiss." The fact that he was of Jewish ancestry caused additional difficulty in being accepted as a true Swiss.

In the midst of this dark period there appeared a bright light. A fellow student of Einstein's at the Polytechnic, Marcel Grossmann, introduced him to a man named Haller, the director of the patent office in Bern. He was a very broad-minded, intelligent man who knew that in every profession it is more important to have someone capable of independent thinking than a person trained in a particular routine. After a long interview he was convinced that although Einstein had no previous experience with technical inventions, he was a suitable person for a position in the patent office, and gave him a job.

In many respects Einstein's removal to Bern was an important turning-point in his life. He now had a position with a fixed annual salary of about three thousand francs, a sum that at that time enabled him to live quite comfortably. He was able to spend his leisure hours, of which he had many, in scientific investigation. He was in a position to think of marriage and of having a family.

Soon after his arrival in Bern, Einstein married Mileva Maritsch, his fellow student at the Polytechnic. She was somewhat older than he. Despite her Greek Orthodox background she was a free-thinker and progressive in her ideas, like most of the Serbian students. By nature she was reserved, and did not possess to any great degree the ability to get into intimate and pleasant contact with her environment. Einstein's very different personality, as manifested in the naturalness of his bearing and the interesting character of his conversation, often made her uneasy. There was something blunt and stern about her character. For Einstein life with her was not always a source of peace and happiness. When he wanted to discuss with her his ideas, which came to him in great abundance, her response was so slight that he was often unable to decide whether or not she was interested. At first, however, he had the pleasure of living his own life with his family. Two sons were born in rapid succession, and the elder was named Albert after his father. Einstein was very happy with his children. He liked to occupy himself with them, to tell them what went on in his mind; and he observed their reactions with great interest and pleasure.

Einstein's work at the patent office was by no means uninteresting. His job was to make a preliminary examination of the reported inventions. Most inventors are dilettantes, and many professionals are likewise unable to express their thoughts clearly. It was the function of the patent office to provide legal protection for inventions and inventors, and there had to be clearly formulated statements explaining the essential feature of each invention. Einstein had to put the applications for patents, which were frequently vaguely written, into a clearly defined form. He had to be able, above all, to pick out the basic ideas of the inventions from the descriptions. This was frequently not easy and it gave Einstein an opportunity to study thoroughly many ideas that appeared new and interesting. Perhaps it was this work that developed his unusual faculty of immediately grasping the chief consequence of every hypothesis presented, a faculty that has aroused admiration in so many people who have had an opportunity to observe him in scientific discussion.

This occupation with inventions also kept awake in Einstein

an interest in the construction of scientific apparatus. There still exists an apparatus for measuring small electrical charges that he invented at this time. Such work was for him a kind of recreation from his abstract theoretical investigation in much the same way as chess and detective stories serve to relax other scientists. Quite a few mathematicians find amusement in the solution of chess problems and not in some sport or in the movies, and it may well be that a mathematical mind finds the best relaxation by occupying itself with problems that are not to be taken seriously but still require a modicum of logical thinking. Einstein does not like chess or detective stories, but he does like to think up all sorts of technical instruments and to discuss them with friends. Thus even today he is often in the company of his friend Dr. Bucky of New York, a well-known physician and specialist in the construction of X-ray machines, and together they have devised a mechanism for regulating automatically the exposure time of a photographic film depending on the illumination on it. Einstein's interest in such inventions depends not on its practical utility but on getting at the trick of the thing.

II

CONCEPTIONS OF THE PHYSICAL WORLD

BEFORE EINSTEIN

1. *Philosophical Conception of Nature*

The philosophical conception of nature that prevails in any given period always has a profound influence upon the development of physical science in that period. Throughout its history natural science has been cultivated according to two very different points of view. The one viewpoint, which may be called "scientific," has attempted to develop a system with which observed facts could be correlated and from which useful information could be obtained, while the other, which may be called "philosophical," has attempted to explain natural phenomena in terms of a specific historically sanctioned mode of exposition. This difference can best be illustrated in the theory of the motion of celestial bodies. In the sixteenth century the Copernican theory, which maintained that the earth moved around the sun, was useful in the correlation of the position of stars, but it was not considered "philosophically true," since this idea contradicted the philosophical conception of that time according to which the earth was at rest in the center of the universe.

The philosophical conception itself, in the history of science, has suffered changes following revolutionary discoveries. Two main periods are outstanding. In the Middle Ages the understanding of natural phenomena was sought in terms of analogies with the behavior of animals and human beings. For instance, the motions of heavenly bodies and projectiles were described in terms of the action of living creatures. Let us call this view the *organismic* conception. The far-reaching investigations in mechanics by Galileo and Newton in the seventeenth century caused the first great revolution in physical thought and originated the conception of the *mechanistic* view in which all phenomena were explained in terms of such simple machines as levers and wheels. This view enjoyed great success, and because of this, mechanics became the model for all the natural sciences — indeed, for all science in general. It reached its acme about 1870, and then, with increasing discoveries in new fields of physics, there began a process of disintegration. Then in 1905, with the publication of Einstein's first paper on the theory of relativity, began the second great revolution. Just as Newton was instrumental in causing the transition from organismic to mechanistic physics, so Einstein followed with the change from the mechanistic to what is sometimes called the mathematical description of nature.

In order to obtain a good understanding of Einstein's work and a comprehension of the paradoxical fate of his theories, it is necessary to appreciate the great emotional disturbances and the interference of political, religious, and social forces that have accompanied the revolutions in the philosophical conceptions of nature. Just as the Roman Inquisition characterized and condemned the investigations of Copernicus and Galileo as "philosophically false" because they did not fit into its conception of nature, many philosophers and physicists all over the world rejected Einstein's theory of relativity since they could not understand it from their mechanistic point of view. In both cases the reason for the condemnation was not a difference of opinion in the judgment of observations, but the fact that the new theory did not employ the analogies required by the traditional philosophy.

It is certainly true that this rigid insistence on the retention of a specific explanatory analogy has in some cases discouraged the discovery of new laws that would account for newly discovered facts. But it would be a great historical injustice to maintain that this conservatism has always been harmful to the progress of science. The application of a specific conception was an important instrument for the unification of the various branches of science. According to the organismic view, there was no real gap between animate and inanimate nature; both were subject to the same laws. The same situation existed in the mechanistic view, in which living organisms were described in terms of mechanics. Furthermore, the thorough application of an analogy frequently demanded a formal simplification, since it favored theories that derived all experimental evidence from a few simple principles.

Since all of us absorbed the mechanistic conception of nature in our training in school, it has become so familiar to us that we regard it as a triviality. When a theory seems trivial, however, we no longer understand its salient point. Consequently, in order to comprehend the great revolutionary significance that this theory possessed when it first appeared, we must try to imagine ourselves in that period. We shall see that mechanistic science in its early stages appeared as incomprehensible and paradoxical to many people as Einstein's theory does today.

2. *Organismic Physics of the Middle Ages*

When we observe a person's action we find that he is sometimes understandable and at other times incomprehensible. When we see a man suddenly dashing off in a particular direction, it appears strange at first, but when we learn that in that direction gold coins are being distributed gratis, his action becomes understandable. We cannot understand his action until we know his purpose. Exactly the same is true of animal behavior. When a hare rushes off in a hurry, we understand this action if we know that there is a dog after it. The purpose of any motion is to reach a point that is somehow better adapted than the point from which it set out.

Just as different kinds of behavior are exhibited by various organisms depending on their nature, so "organismic science" interpreted the movements executed by inanimate objects. The falling of a stone and the rising of flame may be interpreted as follows: Just as a mouse has its hole in the ground while an eagle nests on a mountain crag, so a stone has its proper place on the earth while a flame has its up above on one of the spheres that revolve around the earth. Each body has its natural position, where it ought to be in accordance with its nature. If a body is removed from this position, it executes a violent motion and seeks to return there as quickly as possible. A stone thrown up in the air tends to return as fast as possible to its position as close as it can get to the center of the earth, just as a mouse that has been driven from its hole tries to return there as soon as possible when the animal from which it fled is gone.

It is of course possible that the stone will be prevented from falling. This occurs when a "violent" force acts on it. According to the ancient philosophers: "A physician seeks to cure, but obstacles can prevent him from achieving his aim." This analogy presents the organismic point of view in probably the crudest form.

There are also motions that apparently serve no purpose. They do not tend toward any goal, but simply repeat themselves. Such are the movements of the celestial bodies, and they were therefore regarded as spiritual beings of a much higher nature. Just as it was the nature of the lower organism to strive toward a goal and flee from danger, so it was the nature of the spiritual bodies to carry out eternally identical movements.

This organismic conception had its basis in teachings of the Greek philosopher Aristotle. Although it was basically a heathen philosophy, it is to be found throughout the entire medieval period with only slight modifications in the doctrine of the leading Catholic philosopher, Thomas Aquinas, as well as in the teachings of the Jewish philosopher Moses Maimonides, and the Mohammedan Averroës.

---

3. *Mechanistic Physics and Philosophy*

The transition from organismic to mechanistic physics is most clearly and in a certain sense most dramatically embodied in the person of Galileo Galilei. He looked upon the Copernican theory of the earth's motion as something more than just an "astronomical" hypothesis for the simple representation of observations which says nothing about reality. He dared to throw doubt on the very basic principle of medieval physics.

Galileo took as his starting-point the motion of an object along a straight line with constant velocity. This is a type of motion that is most easily represented by a mathematical treatment. He then considered the motion along a straight line with constant acceleration; that is, when the velocity increases by a constant amount during each unit of time. Galileo tried to understand more complex types of motion on the basis of these simple forms. In particular he discovered as a characteristic property of all falling bodies and flying projectiles that their downward acceleration was constant. He was thus able to consider their entire motion as being made up of two components:

(1) a motion where the initial velocity remains constant both in direction and in magnitude (inertial motion); and

(2) a motion with constant acceleration directed vertically downward (action of gravity).

Sir Isaac Newton later extended this scheme to the more complicated motion of the celestial bodies and then to all motion in general. For the circular motion of the planets, such as the earth, around the sun, Newton decomposed the motion into:

(1) the inertial motion, where the initial velocity remains constant both in direction and in magnitude; and

(2) the action of the gravitational force between the sun and the earth whereby the earth receives an acceleration that is directed toward the sun and is inversely proportional to the square of the distance between the earth and the sun.

He then developed these ideas into his celebrated laws of motion and the theory of gravitation:

Law 1: Every body continues in its state of rest or of uniform motion in a straight line unless it is compelled to change that state by forces impressed upon it (Law of Inertia);

Law 2: The change of motion is proportional to the force impressed, and takes place in the direction in which the force is impressed (Law of Force);

Law 3: To every action there is always opposed an equal reaction; and

The Universal Law of Gravitation: Every particle of matter in the universe attracts every other particle with a force whose direction is that of the line joining the two, and whose magnitude is directly proportional to the product of their masses and inversely proportional to the square of the distance from each other.

The remarkable success of these laws is too well known to need amplification. They have formed the basis for all physics, astronomy and mechanical engineering.

Newton and his contemporaries had already advanced theories concerning optical phenomena. All these theories had one feature in common: they assumed that the laws of mechanics which have been found so successful in calculating the motions of the heavenly bodies and of the material bodies encountered in daily life could be applied also to optical phenomena, and attempted to explain them in terms of motions of particles. Very similar attempts were also made for all processes in other branches of science; for instance, electromagnetism, heat, and chemical reactions. In each case the particular phenomenon was explained in terms of a mechanical model that obeyed the Newtonian laws of motion.

The great practical successes of this method soon reached a point where only an exposition based on a mechanical analogy was considered as giving a satisfactory "physical understanding."

Any other means of presenting and calculating a series of phenomena may be "practically useful," but does not permit a "physical understanding." Explanations in terms of mechanical processes soon began to play the role that explanations in terms of organismic physics had played during the Middle Ages. A mechanistic philosophy took the place of organismic philosophy.

Yet it is obvious that, originally, mechanistic physics owed its success only to its practical utility and not to any kind of philosophical plausibility. The law of inertia when it was first advanced was not plausible from the point of view of the dominant medieval philosophy; on the contrary, it was absurd. Why should an ordinary terrestrial body move along a straight line and forever strive to attain infinity, where it has no business? Yet this "absurd" law overcame all opposition; in the first place because it was mathematically simple, and in the second because the mechanistic physics based upon it led to great successes. Eventually the entire development was turned upside down and it was asserted that only explanations in terms of a mechanical model were "philosophically true." The philosophers of the mechanistic period, especially from the end of the eighteenth century on, excogitated all kinds of ideas to prove not only that the law of inertia was not absurd, but that its truth was evident simply on the basis of reason and that any other assumption was inconsistent with philosophy.

Therein lies the historical root of the struggles waged by many professional philosophers against Einstein's theories. Allied with them were also many experimental physicists whose outlook on more general problems had not grown up on the basis of the scientific principles that they used in their laboratories. They kept their scientific investigations separated from the traditional philosophy that they had learned in the universities and in which they believed as in a creed rather than as in a scientific theory.

4. Relativity Principle in Newtonian Mechanics

There was one point, however, in Newton's laws of motion that was not clear. And this point is very important. The law of inertia states that every body moves in a straight line with constant velocity unless compelled by external influences to change that state. But what is the meaning of the expression "moves in a straight line"? In daily life it is quite clear; when a billiard ball moves parallel to an edge of a table it moves in a straight line. But the table rests on the earth, which rotates about the polar axis and also revolves about the sun. To someone outside the earth the same ball would seem to move in a very complicated path. Hence the ball apparently moves in a straight line only relative to a person in the same room.

Newton explained this point by defining "absolute motion," as "translations of a body from one absolute position to another," and then saying "'absolute' motion is neither generated nor altered, but by some force impressed upon the body moved." Thus if we observe a ball moving parallel to an edge of a table without any force acting on it, then the room can be regarded as resting in "absolute space." Such "resting" rooms in which the law of inertia holds were later called *inertial systems*. If a room, say on a merry-go-round, rotates relative to the "resting" room, then a ball cannot move parallel to an edge of a table standing on the carousel without the exertion of some force. A merry-go-round is no inertial system.

But what if the second room performs a uniform motion in a straight line? The ball can then travel parallel to an edge without the exertion of any force in the "moving" room. In fact, all motion that occurs with uniform velocity in a straight line in the first system will also take place with uniform velocity in a straight line in the second system. Consequently the law of inertia also holds in the "moving system," and it is true whatever the velocity of the system is with respect to the "resting" room as long as it takes place in the same direction with constant magnitude.

When forces operate on the ball and its velocity does not remain constant but acceleration is introduced, the acceleration will be the same in both systems. Hence the law of force, which determines acceleration and is independent of the initial velocity, is the same for both systems. Thus we cannot determine the velocity with which the room moves in relation to the original inertial system by means of experiments on the motion of particles performed in this room; and conversely, with the law of force and the initial velocities, we can predict the future motions of particles without knowing anything about the uniform velocity with which the room may be moving. All systems that move uniformly along a straight line relative to an inertial system are likewise inertial systems. But Newton's laws do not say which material body is an inertial system.

For most practical purposes, the effect of the rotation and revolution of the earth is very small, and its motion can be regarded as a uniform motion in a straight line. Within this range the earth is approximately an inertial system and we can predict the motion of particles on the earth by means of Newton's laws. The same can be done on railway trains, in elevators, and in ships as long as their motion relative to the earth is in a straight line with constant velocity. It is a common experience that we can play with a ball in exactly the same way whether we are on board a train or in a ship so long as it does not jerk or roll.

This law concerning the possibility of predicting future motions from the initial velocities and the laws of force may be called the *relativity principle of mechanistic physics*. It is a deduction from the Newtonian laws of motion and deals only with relative motions and not, as Newton's laws proper, with absolute motion. In this form it is a positive assertion, but it can also be formulated in a negative way, thus: It is impossible by means of experiments such as those described above to differentiate one inertial system from another.

Thus the relativity principle first appeared as a characteristic feature of Newtonian mechanics. As we shall see, it was Einstein's greatest achievement to have discovered that this principle still applies even when Newtonian mechanics is no longer valid. He saw that the relativity principle is more suitable than the Newtonian laws to serve as a basis for a general theory of physical phenomena. It continues to remain valid even when mechanistic physics becomes untenable.

---

5. *Ether as a Mechanical Hypothesis*

The explanation of optical phenomena such as reflection and refraction of light was first given in terms of two opposing theories. Newton had propounded the corpuscular theory, in which light was considered as a stream of particles that behaved according to the laws of motion, while a contemporary of his named Huygens had proposed the wave theory, in which light was considered as a vibration in a certain medium in the same way that sound is a vibration in air. The controversy was settled about 1850 in favor of the wave theory by the French physicists Arago and Foucault. Then the theoretical calculations of Maxwell and the experimental work of Hertz established the result that these vibrations associated with light are electromagnetic in nature; that is, that light is due to very rapid oscillations of electric and magnetic fields.

These vibrations which give rise to propagation of waves require a certain medium in which to oscillate. Sound is due to vibrations of the molecules in the air; there is no sound in a vacuum. Seismic waves, by which earthquakes are recorded, are due to the vibrations of the interior matter of the earth. Water waves are due to the motion of the water on the surface. But light from distant stars reaches us even though there is apparently no material medium in interstellar space. Nevertheless, according to mechanistic physics, it is absolutely essential that the oscillations that give rise to propagation of light have some medium in which to oscillate. This medium was called the *ether*.

Two questions arise when we consider the analogy between sound waves in air and light waves in ether. When any object such as an airplane or a projectile moves through air, there is a certain resisting force due to the friction, and a certain amount of air is dragged along with the object in its progress through the air. Hence the first question: Is it possible to detect motion of objects through the ether, say that of the earth as it revolves around the sun? And the second: Does the ether impede the progress of objects that move through it, and is there any dragging effect?

In order to answer these questions it is necessary to consider the properties of the propagation of light through the ether, since it is only by means of light that ether manifests itself. Now, if a flash of light is just like the spread of ripples on a stagnant pond, its velocity of propagation will have a fixed value with respect to the ether; and to any observer who is moving with respect to it, the velocity will be greater or less depending on whether the direction of the propagation and the motion of the observer are in opposite directions or in the same direction. Thus if the earth moves through the ether without dragging it along in its revolutions around the sun, its velocity relative to the ether should be observable by measuring the velocity of light relative to the earth in different directions.

The fact that the earth moves through the ether without affecting it is known by the *aberration* of starlight. The way in which the spread of light from a star is seen by an observer on the earth, which revolves around the sun, is like that in which a person watches a performance on a stage from a platform that revolves around it. It will appear to him that everything on the stage exhibits periodic annual changes. Astronomers have long known that the fixed stars undergo such annual apparent motions. Thus the phenomenon of aberration shows that the ether is not influenced by the motion of the earth.

The decisive experiment to find the relative motion of the earth through the ether was first prepared at the United States Naval Academy in 1879 by A. A. Michelson. It was carried out afterward at the Astrophysical Observatory in Potsdam, where he spent a year of research, and repeated later in the United States. Michelson, who was the outstanding expert on precise optical measurements, had arranged the experimental conditions so that a definite measurement could be made even if the velocity of the earth through the ether were only a small fraction of that due to its revolution around the sun. The result, however, was entirely negative. It was impossible to find any relative motion of the earth through the ether.

Thus the mechanistic theory of light led to a dilemma. The *aberration* showed that the earth moved through the ether without disturbing it, but the *Michelson experiment* showed that it was not possible to find the velocity with which the earth traveled through the ether.

---

6. *Remnants of Medieval Concepts in Mechanistic Physics*

In medieval physics the characteristic feature concerning the motion of objects had been the revolution of the heavenly bodies around the earth taken as the fixed center. This system represented a kind of a universal framework within which everything had its proper place, and motion within this system meant motion relative to this framework. The problem of absolute motion hardly appeared. Also a natural measure of time was given by the period of revolutions of the heavenly bodies.

It may seem at first that the Copernican theory and the mechanics of Galileo and Newton had disrupted this "closed world" of the Middle Ages, but a careful examination shows that a similar concept was still retained in mechanistic physics. Newton's law of inertia implied that freely moving objects can travel beyond all spatial limits, but it was in relation to "absolute space." Since the connection between absolute space and the empirical content of physical laws was difficult to demonstrate, the auxiliary concept of "inertial system" was introduced. It was not possible, however, to explain why the law of inertia should be valid in certain systems and not in others. This characteristic was not related to any other physical property of the system. Thus the inertial system still retained something of the character of the medieval universal framework. Furthermore, in extending the laws of mechanics to optical phenomena, it had been found necessary to "materialize" space with ether. This ether was a genuine universal framework. The motion of a laboratory relative to it should be observable by means of optical experiments.

The physicists of the mechanistic period always felt uneasy in using the expressions "absolute space," "absolute time," "absolute motion," "inertial system," and "universal ether." Newton himself did not succeed in explaining how one recognized the motion of a body in "absolute space" by actual observation, and he wrote: "It is indeed a matter of great difficulty to discover, and effectually to distinguish, the true motion of particular bodies from the apparent; because the parts of that immovable space, in which those motions are performed, do by no means come under the observation of our senses." Consequently, if one remains within the bounds of physics, one cannot give a satisfactory definition of "absolute motion." The theory becomes completely and logically unobjectionable only if, as was self-evident for Newton, God and his consciousness are added to the physical facts.

For a long time no one had realized precisely what was the actual link between Newton's theological reflections and his scientific work. It was often asserted that they had no logical connection and that his reflections were significant only from a purely emotional standpoint or as a concession to the theological spirit of his time. But this is certainly not so. Although there might have been some doubt about this point earlier, yet since the discovery of the diary of David Gregory, a friend and student of Newton's, we know definitely that Newton introduced the theological hypothesis in order to give his theory of empty and absolute space a logically unobjectionable form. Gregory's diary for 1705 contains an entry concerning a conversation with Newton on this topic. It says: "What the space that is empty of body is filled with, the plain truth is that he [Newton] believes God to be omnipresent in the literal sense; and that as we are sensible of objects when their images are brought home within the brain, so God must be sensible of everything, being intimately present with everything: for he [Newton] supposes that as God

35

is present in space where there is no body, he is present in space when a body is also present."

E. A. Burtt in *The Metaphysical Foundations of Modern Physical Science,* published in 1925, interprets correctly:

"Certainly, at least, God must know whether any given motion is absolute or relative. The divine consciousness furnishes the ultimate center of reference for absolute motion. Moreover, the animism in Newton's conception of force plays a part in the premise of the position. God is the ultimate originator of motion. Thus in the last analysis all relative or absolute motion is the resultant of an expenditure of the divine energy. Whenever the divine intelligence is cognizant of such an expenditure, the motion so added to the system of the world must be absolute."

By means of this anthropomorphic conception of God, a scientific, almost physical definition of absolute motion is obtained. It is linked with the energy expended by a being called "God," but to which properties of a physical system are ascribed. Otherwise the concept of energy could not be applied to the system. Fundamentally the definition means that one assumes the existence in the world of a real source of energy that is distinguished from all others. Motion produced by the energy expenditure of mechanical systems in general is described as only "relative" motion, while motion produced by this select being is characterized as "absolute." It should never be forgotten, however, that the logical admissibility of this definition of absolute motion is bound up with the existence of the energy-producing being. During the eighteenth century, in the age of the Enlightenment, men no longer liked to ascribe to God a part in the laws of physics. But it was forgotten that Newton's concept of "absolute motion" was thereby deprived of any content. Burtt in his aforementioned book says very aptly: "When, in the eighteenth century, Newton's conception of the world was gradually shorn of its religious relations, the ultimate justification for absolute space and time as he had portrayed them disappeared and the entities were left empty."

7. *Critics of the Mechanistic Philosophy*

Toward the end of the nineteenth century more and more physical phenomena were discovered that could be explained only with great difficulty and in a very involved way by the principles of Newtonian mechanics. As a consequence new thories appeared in which it was not clear whether they could be derived from Newtonian mechanics, but which were accepted as temporary representations of the observed phenomena. Was this true knowledge of nature or only a "mathematical description," as the Copernican system was considered in medieval physics? These doubts could not be resolved so long as it was believed that there were philosophical proofs according to which reduction to Newtonian mechanics provided the only possibility for the true understanding of nature.

During the last quarter of the nineteenth century a critical attitude toward this mechanistic philosophy became more and more evident. An understanding of this criticism is an essential prerequisite for the understanding of Einstein's theory and its position in the development of our knowledge of nature. As long as it was believed that Newtonian mechanics was based ultimately on human reason and could not be shaken by scientific advance, every attempt such as that of Einstein, to establish a theory of motion not founded on Newton's theory necessarily appeared absurd. The critics of mechanistic philosophy plowed the soil in which Einstein was then able to plant his seeds.

As the first of these critics, we may mention Gustav Kirchhoff, the discoverer of spectral analysis. In 1876 he stated that the task of mechanics was "to describe completely and as simply as possible motions occurring in nature." This meant that Newtonian mechanics is itself only a convenient scheme for a simple presentation of the phenomena of motion that we observe in daily experience. It does not give us an "understanding" of these occurrences in any other philosophical sense. By thus contravening the general opinion that Newton's principles of mechanics are self-evident to the human mind, he created something of a sensation among natural scientists and philosophers.

Furthermore, with Kirchhoff's conception that mechanics is only a description of the phenomena of motion, the mechanical explanations of the phenomena in optics, electricity, heat, etc. — the aim of mechanistic physics — became simply descriptions of these results in terms of a pattern that had been found to be most suitable for mechanics. Why should one describe by this roundabout method of using mechanics instead of trying to find directly the most suitable scheme for the description of various phenomena? Newtonian mechanics was thus deprived of its special philosophical status.

In 1888 Heinrich Hertz discovered the electromagnetic waves, which form the basis of our modern wireless telegraphy and radio, and he then set out to explain these phenomena in terms of a physical theory. He took as his starting-point Maxwell's theory of electromagnetic fields. James Clerk Maxwell had derived his fundamental equation from mechanistic physics by assuming that electromagnetic phenomena are actually mechanical oscillations in the ether. Hertz noticed that in doing this Maxwell had been compelled to invent mechanisms that were very difficult to calculate, and found it was simpler to represent electromagnetic phenomena directly by means of Maxwell's equation between electric and magnetic fields and charges. Since it was also evident to him, however, that these relations could not be derived directly from experience, he was led to a consideration of the logical character of these equations. In 1889 he made a remark that can be regarded as the program for the new approach to physics, a conception that was eventually to replace the mechanistic view. Hertz said:

"But in no way can a direct proof of Maxwell's equations be deduced from experience. It appears most logical, therefore, to regard them independently of the way in which they had been arrived at, and consider them as hypothetical assumptions and let their plausibility depend upon the very large number of natural laws which they embrace. If we take up this point of view we can dispense with a number of auxiliary ideas which render the understanding of Maxwell's theory more difficult."

Thus Hertz consciously abandoned that which during both the organismic and the mechanistic period was described as the "philosophical" foundation of physics. He maintained that it was sufficient to have a knowledge of laws from which phenomena could be calculated and predicted without raising any question of whether these laws were intrinsically evident to the human mind.

---

8. *Ernst Mach: The General Laws of Physics Are Summaries of Observations Organized in Simple Forms*

The criticisms of the mechanistic philosophy by physicists such as Kirchhoff and Hertz were only occasional and aphoristic. There were others, however, whose criticisms were based on a very precise conception of nature and of the task of science. The French philosopher Auguste Comte advanced the sociological theory that the "metaphysical" stage in the development of a science is already succeeded by a "positivistic" one. This means that the demand for the use of a specific analogy such as the organismic and mechanistic views is abandoned and after that a theory is judged only as to whether it presents "positive" experience in a simple, logically unobjectionable form.

This approach was most widely and profoundly developed by the Austrian physicist Ernst Mach, who became one of Einstein's immediate forerunners. Mach carried out a thorough historical, and logical analysis of Newtonian mechanics and showed that it contains no principle that is in any way self-evident to the human mind. All that Newton did was to organize his observations of motion under several simple principles from which movements in individual cases can be predicted. But all these predictions are correct only so long as the experiences upon which Newton based his principles are true.

Mach emphasized, in particular, the demand for simplicity and *economy of thought* in a physical theory: the greatest possible number of observable facts should be organized under the fewest possible principles. Mach compared this requirement to the demand for economy in practical life and spoke of the "economic" nature of scientific theories. Thus Mach, instead of demanding the use of a specified analogy, insisted that science be "economical."

Furthermore, not only did Mach criticize the attempts of philosophers to make a philosophical system out of Newton's mechanics, but he also criticized the remains of medieval physics that it still retained. He pointed out that Newton's theory contained such expressions as "absolute space" and "absolute time," which cannot be defined in terms of observable quantities or processes. In order to eliminate such expressions from the fundamental laws of mechanics, Mach raised the demand which is now frequently described as the *positivistic criterion* of science: namely, that only those propositions should be employed from which statements regarding observable phenomena can be deduced.

This demand is very aptly elucidated by his criticism of Newton's law of inertia. If we wish to test this law experimentally, we can never formulate a question such as this: Does a body tend to maintain the direction of its initial velocity relative to absolute space? The question is meaningless since absolute space is unobservable. If we perform, say, Foucault's pendulum experi-ment, which gives an experimental proof of the rotation of the earth, we observe actually that the pendulum maintains its plane of oscillation relative, not to absolute space, but rather to the fixed stars in the sky.

Consequently, according to Mach, all mention of absolute space should be removed from the law of inertia, and it would then be expressed as follows: Every body maintains its velocity, both in magnitude and in direction, relative to the fixed stars as long as no forces act upon it. This means that the fixed stars exert an observable influence on every moving body, an effect that is in addition to and independent of the law of gravitation. For the motion of terrestrial objects this latter influence is hardly observable in practice, since the force of gravity decreases with the square of the distance between the attracting bodies, but the laws of inertia will determine all terrestrial motion if the frame-work of the fixed stars is declared as an inertial system.

9. *Henri Poincaré: The General Laws of Physics Are Free Creations of the Human Mind*

In consequence of the criticisms of Mach and others, it had become clear that the laws of Newtonian mechanics and the understanding of all physical phenomena in terms of it are not demanded by human reason. However, Mach's assertion that the general laws of physics are only simple economical summaries of observed facts was not satisfactory to many scientists. Particu-larly for physicists who thought along mathematical lines and had a greater formal imagination, the assertion, for example, that Newton's law of gravitation is only a simple summary of observa-tion on the positions of the planets did not seem adequate. Be-tween the actual observation of the position of the planets by a telescope and the statement that the gravitational force between two bodies is inversely proportional to the square of the distance there seemed to be a wide gap.

Criticism of nineteenth-century physics in this direction was carried on chiefly by the French mathematician Henri Poincaré. His writings on the logical character of the general laws of na-ture probably exerted more influence on mathematicians and physicists toward the end of the nineteenth century than any other similar writings. He paved the way for a new, logically satisfying conception of nature, and his ideas also played an outstanding part in the reception and discussion of Einstein's theories.

Poincaré's view is often described as "conventionalism." According to him, the general propositions of science, such as the theorem about the sum of the angles of a triangle, the law of inertia in mechanics, the law of conservation of energy, and so on, are not statements about reality, but arbitrary stipulations about how words, such as "straight lines," "force," "energy," are to be employed in the propositions of geometry, mechanics, and physics. Consequently one can never say whether one of those propositions is true or false; they are free creations of the human mind and one can only question whether these stipulations or conventions have been expedient or not.

This conception may be elucidated by means of two examples. Let us first consider the geometrical theorem referred to above: namely, that the sum of the angles of a triangle is equal to two right angles. According to nineteenth-century tradition this is an unshakable proposition, which is a product of human reasoning and at the same time a statement concerning what is actually observed in nature. On the one hand, we can derive this proposition from the axioms of geometry, which are "directly evident to the mind"; on the other hand, by measuring the angle of an actual material triangle, we can corroborate this relationship. Poincaré, however, says: if an actual triangle is formed from, say, three iron rods, and the measurement shows that the sum of the angles is *not* exactly equal to two right angles, one of two different conclusions can be drawn: either that the geometrical theorem is not valid, or that the rods forming the triangle are not straight lines. We have the two alternatives, and we can never decide by experiments the validity of geometrical theorems. Consequently we can say that the propositions of geometry are arbitrary stipulations or definitions and not statements about empirical facts. They establish under what circumstances we wish to call a rod a "straight line." Thus geometrical theorems are not statements about the nature of space, as it is often expressed, but rather definitions of such words as "straight lines."

According to Poincaré, the laws of mechanics are of somewhat similar character to the propositions of geometry. Let us, therefore, consider the law of inertia as the second example. The possibility of verification of the law rests on our ability to determine whether or not a body moves with uniform velocity in a straight line. As long as we cannot do this, the law of inertia can only be characterized in some such statement as this: "When a body moves without being influenced by forces, we call this state a uniform motion along a straight line." It is simply a definition of the expression "uniform motion in a straight line," or, according to our discussions in sections 3 and 4, a definition of the term "inertial system."

Thus the general principles, such as the theorem about the sum of the angles of a triangle or the law of inertia, do not describe observable phenomena, but are rather definitions of expressions such as "straight line" or "uniform motion along a straight line." One has to add definitions by which one recognizes whether a given rod is straight or the motion of a ball is uniform and along a straight line and which have been named "operational definitions" by P. W. Bridgman. These, together with the physical laws (e.g., the law of inertia), constitute a a system of propositions that can be verified by experience.

One of the chief consequences of this conception is that it makes no sense in science to inquire into the philosophical significance or the "nature" of such physical expressions as "force," "matter," "electric charge," "duration of time," etc. The use of such concepts is always justified if statements permitting experimental verification can be derived from the propositions in which these expressions occur. Apart from this they have no meaning. Because Newtonian mechanics was able to describe very complex phenomena such as the motion of the planets in simple statements with the aid of the words "force" and "mass," these terms have scientific meaning. There is no need to puzzle one's brain over whether "force" can be explained from a "mechanistic" standpoint or "matter" from an "organismic" one. "Force" and "matter" are constructions of the human mind.

---

10. *Positivistic and Pragmatic Movements*

The idea of Mach that the general laws of science are simple summaries of experimental facts, and the idea of Poincaré that they are free creations of the human mind, appear to be diametrically opposed to each other; but when we consider the intellectual currents of the last quarter of the nineteenth century, we can see that they were only two wings of the same intellectual movement, generally known as the *positivistic movement*. It was directed chiefly against the metaphysical foundations of science. The proponents of this view asserted that the validity of the general principles of science cannot be proved by showing that they are in agreement with some eternal philosophical truths, and they set out to investigate how the validity can be judged within science itself. They found that two criteria are possible, an empirical and a logical. In the former the observable facts that follow from the general principles must have experimental confirmation, and in the latter the principles and operational definitions must form a practical and consistent system. The emphasis put on the empirical or the logical criterion determined one's position in one or the other wing of the movement. Mach was on the extreme empirical wing, while Poincaré was on the extreme logical side. There was therefore no conflict between them; it was only that two different aspects of the same scientific method were being emphasized.

The positivistic movement exerted a great influence in central and western Europe during the last quarter of the nineteenth century. The central European positivism, chiefly centered in the Austrian Ernst Mach, was to be found in the universities of Vienna and Prague. It had but little influence and few followers in the universities of the German Reich. At this time Germany was completely under the influence of various versions of Kantian philosophy, whose status was almost that of a state religion. Since German was also the chief language of science in Austria, Central European positivism developed largely as the critic and rival of Kantian philosophy. For this reason it was more militant than French positivism, led by Poincaré.

About this time there appeared independently in the United States a movement that is related to European positivism in its chief line of thought. In 1878 C. S. Peirce published an essay on the logical character of scientific statements. Like Mach and Poincaré, he pointed out that the meaning of general statements cannot be derived from agreement with still more general metaphysical truths, but must be drawn from the observed facts that follow from them. In contrast to the European positivists, however, Peirce emphasized particularly the role of propositions as the basis for our actions. He therefore called his doctrine "pragmatism." "The essence of a belief," he said, "is the establishment of a habit, and different beliefs are distinguished by different modes of action to which they give rise." Like Mach, Peirce also warned against the trivial metaphysics that we have imbibed with our education since childhood. He said: "The truth is that common sense or thought as it first emerges above the level of the narrowly practical is deeply imbued with that bad logical quality to which the epithet metaphysical is commonly applied." He also emphasized that words such as "force" are only expedients for the representation of facts and that every question as to their "actual nature" is superfluous and useless. In the same article he said:

"Whether we ought to say that a force *is* an acceleration or that it *causes* an acceleration is a mere question of propriety of language which has no more to do with the real meaning than the difference between the French idiom *'il fait froid'* and the English equivalent 'it is cold.'"

An approach very similar to that of Mach was manifested by John Dewey in his first scientific article: "The Metaphysical Assumptions of Materialism," published in 1882. Dismissing the opinion that the reduction of all phenomena to the motions of material bodies is an explanation of nature, he said:

"First, it assumes the possibility of ontological knowledge, by which we mean knowledge of being or substance apart from a mere succession of phenomena. . . . Secondly, it assumes the reality of causal nexus and the possibility of real causation. In declaring that matter causes mind it declares that the relation is one of dependency and not one of succession."

The struggle against materialism here is not carried on in the service of an idealistic philosophy, as with the average professors of philosophy in European and American universities, but entirely along the lines of central European positivism, which opposed mechanistic physics on the ground that it is not a sufficiently broad basis of science.

American pragmatism since then has developed into a powerful movement, finding its most characteristic expression in John Dewey and William James. It has devoted itself more to the problems of human life than to the logic of the physical sciences, in contrast to the development of positivism in Europe. Considered from the purely logical point of view, however, the basic tendency was the same on both sides of the Atlantic. The medieval idea of a philosophical explanation in contrast to a practical representation of facts vital to life lost prestige to an ever increasing degree. From a logical basis of science, metaphysics developed into a means of satisfying emotional needs.

11. *Science at the End of the Nineteenth Century*

During the golden age of mechanistic physics it was generally felt that outside of its application lay the realm of the unknowable and the unintelligible, since "to understand" meant "to represent by analogy to a mechanism." In 1872 the German scientist Du Bois-Reymond, in his famous lecture on *"Die Grenzen des Naturerkennens* (The Limits of Our Knowledge of Nature)" took as his point of departure the assertion, then regarded as self-evident, that "understanding" means "reduction to the laws of Newtonian mechanics." He indicated two important problems of science that can certainly not be reduced to mechanics. These are, first, the problem of what "actually occurs in space where a force is acting" and, secondly, how it happens that "matter in the human brain can think and feel." Since the answers to these questions can obviously not be obtained within the framework of mechanistic physics, he concluded that there are "insoluble problems" that are inaccessible to human knowledge. To these questions we should say *"ignorabimus"* ("we shall never know") instead of *"ignoramus"* ("we do not know"). This word *"ignorabimus"* became the slogan of an entire period, the slogan of defeatism in science, which delighted all anti-scientific tendencies of the period. Toward the end of the nineteenth century more and more facts became known in physics and biology that could not be explained or controlled by means of the laws of mechanics, with the result that the catchword *"ignorabimus"* was soon converted into the even more exciting slogan, "the bankruptcy of science."

This feeling of the failure of rational scientific thought was intensified by various social developments. Science — that is, science guided by the spirit of mechanistic physics — had led men during the eighteenth and nineteenth centuries to believe in the possibility of continual progress. If men only acted according to the teachings of science instead of irrational superstitions, mankind would be freed from all need. The political expression of this faith was liberalism. Toward the end of the nineteenth century, however, it became ever clearer that the attempts based on science and the faith in progress had not succeeded in abolishing the economic misery of the great mass of the population, or in eliminating the psychological suffering of individual human beings. A feeling of despair developed which expressed the conviction that scientific theory and practice were a disappointment. Alongside liberalism, new political currents developed that had their own conceptions of science, conceptions differing from the mechanistic view. One tendency propagated a return to the organismic science of the Middle Ages, and from it developed the authoritarian socialism that became the germ-cell of later fascism in all its varieties. Another movement, represented by Karl Marx, wanted to transform "mechanistic" materialism into "dialetical" materialism, and from it developed the communism of the twentieth century.

It was impossible to deny that science was still the basis of technological progress, but it was believed that it could be disparaged by speaking of it as the church did about the Copernican system of the world: that mechanistic natural science provided only a useful guide for action, and no true knowledge of nature. Around 1900 Abel Rey, a French philosopher and historian of science, gave a very acute and trenchant description of the dangers for general intellectual life entailed by such an attitude of despair. He said:

"If these sciences which have had an essentially emancipating effect in history go down in a crisis which leaves them only with the significance of technically useful information but robs them of every value in connection with the cognition of nature, this must bring about a complete revolution. The emancipation of the mind as we owe it to physics is a most fatally erroneous idea. One must introduce another way, and give credit to subjective intuition, to a mystical sense of reality."

There have been two ways out of this crisis of science which had developed in consequence of the breakdown of mechanistic physics. In his book *The Idealistic Reaction against Science,* Aliotta, an Italian, described the situation in the following very striking manner:

"Could thought rest easy in this complacent agnosticism? There were two ways of escaping this intolerable situation: either to turn to the other function of the mind (besides intellect) or to eliminate the problem altogether by proving that it is due to a faulty perspective and false conception of science. Both ways have been tried. On the one hand, by a return to the moralism of Fichte and the æstheticism of the romanticists, into which the rebellious genius of Nietzsche has breathed new life, the will as the creative source of all values and of unfettered æsthetic intuition is exalted above intelligence. On the other hand, the bases of the mechanical conception and its chief instruments — geometrical intuition and mathematical calculation —

are subjected to a searching examination. This analysis, to which men of science themselves were impelled by the discovery of the new principle of energy and by metageometrical concepts, resulted in stress being laid upon the active work of the mind in the construction of scientific laws and theories."

The second alternative mentioned here was the view taken by the advocates of positivism and pragmatism. Their way out of the "bankruptcy of science" was to proclaim that mechanistic science had formulated the problem in a manner that necessarily led into a cul-de-sac; it had not correctly defined the goal of science. The unattainable something, for which the despairing solution of *"ignorabimus"* was proposed, had been recognized as a phantom, a chimera that has nothing to do with science. Through an analysis of the really successful methods of science, men such as Mach and Poincaré in Europe, and Peirce and Dewey in America, have shown that it is of no significance whatsoever whether observations are presented in terms of a certain preferred analogy. All that matters is that the statements of science are useful; the specific language and the equation by which they are formulated do not matter. Thus with the goal of science defined in the positivistic and pragmatic sense, it becomes evident that the end of the nineteenth century does not represent a crisis, but rather one phase of the gradual progress of science toward its goal, which is the creation of an instrument for predicting and controlling the phenomena.

In a certain sense this positivistic-pragmatic movement, so characteristic of the turn of the century, belonged to the group of movements that were directed against the overestimation of the role of the intellect. Professor Ralph Barton Perry very correctly said:

"Much the most sophisticated form of anti-intellectualism and at the same time the form most characteristic of our age is that form which has now come very generally to be called 'instrumentalism' and which is represented at present by the school of James and Dewey in America. . . . According to this view the intellect instead of being an oracle is a practical instrument to be judged by the success with which it does work."

Nevertheless, the new movement, no matter whether it was called pragmatism, positivism, or instrumentalism, could be characterized as anti-intellectual only in so far as it warned against occupying the intellect with meaningless problems. The adherents said that the intellect is unable to discover the metaphysical reality behind phenomena. But this is not a diminution of its role, since to speak of such a metaphysical reality does not make sense for science. It is sterile and leads only to confusion. The creation of an "instrument," which is now what is meant by "science," can be accomplished only by means of the intellect, even though we cannot produce a blueprint for discovering general principles. The discovery of laws such as the energy principle or the law of inertia is the work of a genius, like the composition of a symphony. But when the general law has been enunciated, it is then the function of the methodically proceeding intellect to make its meaning clear to all. Only the intellect can test the principle and pronounce judgment on its truth — that is, whether it is of value in realizing the aims of science.

So ends the nineteenth century. Its faith in the ability of science to reveal the ultimate reality behind phenomena was shaken; but in its place appeared the sober consolation of positivism that science had become more flexible and girded for new tasks of a boldness never dreamed of. During the twilight period characterized by a devaluation of the intellect and an increased regard for action, there appeared, like a silver glow on the horizon, the hope that a more acute logical analysis would give an entirely new form of science based on a methodically operating intellect. The twentieth century ushered in this dawn.

III

BEGINNING OF A NEW ERA IN PHYSICS

1. *Life in Bern*

When Einstein took up his position at the patent office in Bern, it was in two respects a turning-point in his life. He became engaged in a practical occupation that made him financially independent and filled his time with an obligatory activity, and he founded a family. For most people these two circumstances provide the most important and often the only content of their lives. This was true only to a very slight degree for Einstein, to whom neither professional activity nor a family had a great significance. At times these activities gave him a certain relaxation, but they never really satisfied him.

Throughout his life Einstein has been in a certain sense a very lonesome man. He sought the harmony of the universe in music as well as in mathematical physics and he has been engaged in these two fields during his entire life. Everything else was significant for him only in so far as it affected his progress toward this goal. He sought friends with whom he could play music or discuss ideas about the universe; yet he did not like to become so intimate with his friends that they could in any way interfere with his freedom. His attractive, frank, and witty personality easily made many friends, but his predilection for isolation and his concentration on his artistic and scientific life disappointed many people and estranged some who had been, or at least had believed themselves to be, his friends. We find repeatedly that throughout his life this contrast has determined his relations to his environment.

Much later (1930) he himself described this character trait very precisely and strikingly:

"My passionate interest in social justice and social responsibility has always stood in curious contrast to a marked lack of desire for direct association with men and women. I am a horse for single harness, not cut out for tandem or teamwork. I have never belonged wholeheartedly to any country or state, to my circle of friends, or even to my own family. These ties have always been accompanied by a vague aloofness, and the wish to withdraw into myself increases with the years. Such isolation is sometimes bitter, but I do not regret being cut off from the understanding and sympathy of other men. I lose something by it, to be sure, but I am compensated for it in being rendered independent of the customs, opinions, and prejudices of others, and am not tempted to rest my peace of mind upon such shifting foundations."

Although Einstein did not seek much stimulation from others, yet he did not like to develop his ideas in solitude without any contact with other people. Frequently he has liked the presence of a companion in order to be able to speak his mind freely. Even during the early period in his career he liked to try out his ideas on others to see how they reacted to them.

At Bern his chief companion in this respect was an Italian engineer named Besso. He was somewhat older than Einstein, and a man of critical mind and a highly nervous temperament. He was often able to offer pertinent critical remarks on Einstein's formulations, and also responded vigorously to those ideas of Einstein's which were new and astonishing. He frequently remarked about new ideas: "If they are roses, they will bloom." Around Einstein and Besso there gathered a small group of people interested in science and philosophy, who often met to discuss such questions.

2. *Interest in Philosophy*

Since Einstein was chiefly interested in the general laws of physics or, more precisely, in deriving logically the immeasurable field of our experiences from a few principles, he soon came into contact with a set of problems that are usually dealt with in philosophical works. Unlike the average specialist, he did not stop to inquire whether a problem belonged to his field or whether its solution could be left to the philosophers.

Einstein read philosophical works from two points of view, which were sometimes mutually exclusive. He read some authors because he was actually able to learn from them something about the nature of general scientific statements, particularly about their logical connection with the laws through which we express direct observations. These philosophers were chiefly David Hume, Ernst Mach, Henri Poincaré, and, to a certain degree,

Einstein and his graduating class at the Cantonal School, Aarau, Switzerland, 1896

Einstein and Mileva, his first wife

Einstein in 1905, the year during which he propounded the theory of relativity and the hypothesis of photons (light quanta)

Immanuel Kant. Kant, however, brings us to the second point of view. Einstein liked to read some philosophers because they made more or less superficial and obscure statements in beautiful language about all sorts of things, statements that often aroused an emotion like beautiful music and gave rise to reveries and meditations on the world. Schopenhauer was pre-eminently a writer of this kind, and Einstein liked to read him without in any way taking his views seriously. In the same category he also included philosophers like Nietzsche. Einstein read these men, as he sometimes put it, for "edification," just as other people listen to sermons.

The philosopher whose views Einstein felt helped him most was David Hume, who is usually characterized as the "representative of the English Enlightenment." What Einstein liked most about Hume was the unsurpassable clarity of his presentation and his avoidance of any ambiguities intended to give an impression of profundity. Hume showed that there are only two methods available for science: experience and mathematical-logical derivations. He was the father of the logical-empirical approach, and he rejected all metaphysical auxiliary concepts if they could not be established by experience and logical derivation. The most famous examples are Hume's criticism of the ordinary conception of the relation between cause and effect, and of induction — the method of deriving a general law from a few particular instances.

When we observe that a stone A strikes a stone B and sets it in motion, we usually express this occurrence as follows: Stone A has caused stone B to move. By experience we can only confirm the fact that whenever A strikes B, B is set in motion. Before Hume it was usually said that this connection is a *necessary* one. In physics, however, the word "necessary" can have no meaning other than "regularly connected." If in addition to this we wish to introduce the word "necessary" as "cause" in another, higher sense, we are asserting something that cannot be proved by any observation. Every observation shows only whether or not the motion of B regularly follows when it is struck by A, but never anything that can be expressed by the statement: "Motion of B *necessarily* follows from collision with A."

According to Hume, then, to explain a phenomenon causally means only to state the conditions under which it occurs. This conclusion of Hume's that science knows only the regularity of natural phenomena and processes, but nothing about any "causation" that goes beyond this, was of the greatest significance for

Einstein's scientific thought. Many of the polemics later directed against Einstein were fundamentally polemics against Hume. We shall see that his adherence to the "philosophy of the English Enlightenment" was later used by German nationalists to discredit him. It was used to tie up Einstein's theories with the political philosophy of liberalism, and consequently to condemn them.

Some of Hume's ideas also appear in the writings of Ernst Mach, the leader of central European positivism. Next to Hume, Mach was the philosopher who exerted the greatest influence on Einstein. Of particular significance was Mach's criticism of the remains of medieval physics in Newtonian mechanics, which have already been discussed in Section 8 of the last chapter. Mach's criticism that such expressions as "absolute space," "absolute time," and "absolute motion" could not be connected in any way with physical observations was one of the points from which Einstein set out to replace Newton's theory of motion by his own. "Mach's postulate" has in many instances been a useful point of departure for new theories. According to this "postulate," for every physical phenomenon the conditions of its occurrence must be sought among other observable phenomena. Later "Mach's postulate" led Einstein to advance his new theory of gravitation.

On the other hand, Einstein was not particularly sympathetic to what he called the "Machian philosophy," by which he meant Mach's doctrine that the general laws of physics are only summaries of experimental results. Einstein believed that this conception did not give sufficient credit to the fact that general laws cannot be inferred from experience. In Einstein's opinion they are to be tested by experience, but owe their origin to the inventive faculty of the human mind.

It was this very point that Einstein esteemed so highly in Kant's work. Kant's principal point was that the general laws of science contain not only the result of experience, but also an element provided by human reason. On the other hand, Einstein did not share Kant's belief that human reason by itself can yield important natural laws, and that consequently there are laws that are eternally valid. Einstein liked to read Kant because through him he became acquainted with many of Hume's ideas. The views of Einstein and Kant are similar in their emphasis on the role of the human mind. but this similarity is rather emotional than logical.

3. *The Fundamental Hypotheses of the Theory of Relativity*

The blind alley into which the ether theory of light had been led by Michelson's experiment has already been mentioned in Section 5 of the previous chapter. Michelson had tried to measure the velocity of the earth as it moves through the ether, but had obtained the value zero for this velocity.

The main idea of this experiment can be explained in this way: We know that a swimmer takes longer to swim upstream than downstream between two points in the bank. In fact, by measuring the two rates of travel, we can easily calculate the velocity of both the swimmer and the stream. According to the mechanistic view, light should travel through the ether in exactly the same way as the swimmer in the stream, and experiments on light propagated through the "stream of ether" relative to the moving earth should be comparable to observations on the swimmer made from the bank of the stream. Thus the measurements of the velocities of light when traveling with the ether stream and against it should enable us to calculate the velocity of the earth through the ether. The execution of this fundamental idea in this simple form, however, is not practicable, since the velocity of light is so very great — 186,000 miles per second — but Michelson devised a means whereby the velocities of light that has traveled along two well-defined paths could be compared. His idea was to measure the difference in time taken by one beam, which travels from a certain point (S) to a mirror (M) in a direction along the motion of the earth through the ether and then back to S against the motion, and another beam which goes from S to another mirror (N) situated the same distance from S as M, but in a direction perpendicular to the motion and back to S. If the mechanistic view is correct, the first beam should take a slightly longer time than the second, and with the sensitive apparatus that Michelson had, the result should have been observable, even if the velocity of the earth through the ether were only a small fraction of the velocity of the earth around the sun. There was no observable difference in the two times, however.

If we refuse to assume that the earth always remains at rest in the ether, which would contradict other observations, the only possible conclusion that we can draw from Michelson's experiment is that the hypothesis on which the result was predicted must be false. This hypothesis, however, was the mechanistic theory of light itself.

Einstein drew a radical conclusion and suggested abandoning entirely the assumption that light is a process in a medium known as the ether. Instead of asking what are the results of the interaction of light and motion according to the ether theory of light, he asked what are the chief characteristics of the interaction of light and motion that are known from actual observations. He condensed these features into a few simple laws and then inquired what could follow from such laws if developed along logical and mathematical chains.

Michelson's experiment and similar ones performed by others showed that optical phenomena cannot be regarded as mechanical phenomena in the ether, but that they do have a very general observable feature in common with mechanical phenomena. This feature which is common to the motion of material bodies and the propagation of light Einstein found in the principle of relativity.

As we have seen in Section 4 of the last chapter, Newtonian mechanics contained a relativity principle, which stated that the future motion of any object with respect to an inertial system can be predicted from its initial position and velocity with respect to this system, without any knowledge of the motion of the inertial system itself.

Now, if we disregard the existence of the ether, the null result of Michelson's experiment means exactly that the result can be predicted from the experimental arrangement in the laboratory without any knowledge of its speed with respect to the celestial bodies. Since similar statements could be made for other optical phenomena, Einstein proposed to extend the relativity principle of Newtonian mechanics to include optical phenomena in the following form: "The future course of optical phenomena may be predicted from the conditions of the experiment relative to the laboratory in which it is carried out, without knowing the velocity of the laboratory in the universe." Thus, according to Einstein, the connection between mechanical and optical laws is not based on a reduction of optics to mechanics, but rather on the fact that one and the same general law holds for both.

Besides this "principle of relativity," Einstein needed a second principle dealing with the interaction of light and motion. He investigated the influence of the motion of the source of light on the velocity of the light emitted by it. From the standpoint of the ether theory, it is self-evident that it makes no difference whether or not the source of light moves; light considered as mechanical vibration in the ether is propagated with a constant velocity with respect to the ether. This velocity depends only on the elasticity and density of the ether.

Dropping the ether theory of light, Einstein had to reformulate this law into a statement about observable facts. There is one system of reference, F (the fundamental system), with respect to which light is propagated with a specific speed, c. No matter with what velocity the light source moves with respect to the fundamental system (F), the light emitted is propagated with the same specific velocity (c) relative to F. This statement is usually called the "principal of the constancy of the speed of light."

The constancy of the speed of light has been confirmed empirically from the observation in double stars. They are stars of approximately equal masses, which are close together and revolve about each other, and are well known to astronomers. If the velocity of light depended on the velocity of the source, then as the stars revolve, the time taken for light to reach the earth from the member of the pair that is approaching the earth will be shorter than the corresponding time of the light from the receding member. Analysis of the two beams of light has shown that there is no observable effect from the velocity of the source.

---

4. *Consequences of Einstein's Two Hypotheses*

It was a characteristic feature of Einstein's mode of work to deduce from his fundamental principles all the logical consequences to the limit. He showed that from these hypotheses, which appeared quite harmless and plausible, a rigorous deduction led to results that seemed very novel and in part even "incredible." From these results he went on to others, which not only seemed incredible but were even pronounced "paradoxical," "absurd," and "incompatible with sound logic and psychology."

There are at present thousands of papers in which attempts are made to explain Einstein's theory to the lay public. It is not the purpose of this book to go into all the details of his theories, but to give a description of Einstein's personality and his relation to his environment. It is necessary, however, to go into his scientific work to a certain extent in order to give the reader some idea of the manner in which he attacked scientific problems in comparison with that of other scientists. In particular we should try to understand how it happened that his theories not only were of interest to physicists, but also stimulated and excited philosophers, thus indirectly stirring up a public that had only slight interest in scientific questions but that participated in the general intellectual life of our period.

From the two basic assumptions Einstein was able to conclude not only that the mechanistic theory of light was erroneous, but that even the Newtonian mechanics of material objects could not be generally valid. This result can be fairly easily understood if we trace it back to the way Einstein speculated on the properties of light as early as at the age of sixteen.

While still a student, Einstein had pictured to himself the remarkable things that would occur if a body could travel with the speed of light — at the rate of 186,000 miles per second. Let us consider the fundamental system (F) and a laboratory (L) for optical experiments which moves with constant velocity (v) with respect to F. Let there be a source of light (R) at rest in F, from which a beam of light is propagated with velocity c in the same direction as the laboratory (L) is moving. Now, if the velocity (v) of the laboratory (L) is equal to the velocity of light (c) then, according to Newtonian mechanics, the ray of light will be stationary with respect to the laboratory. No vibration is registered in L. Since light does not move with respect to L, there are no rays in L, and the usual experiments of reflection and refraction cannot be performed (fig. 1).

It is, of course, imaginable that in such a rapidly moving system (L) there would no longer be any optical phenomena in the ordinary sense. This occurrence, however, would be inconsistent with Einstein's principle of relativity in optics. For according to this principle, all optical experiments should give the same result whatever the speed (v) of the laboratory may be.

The same difficulty appears if we compare directly the results of Einstein's two principles (relativity and constancy) within the ether theory of light. We consider again a laboratory that is moving with the speed of light (c) relative to the fundamental system (F). Suppose a mirror is set up in L to reflect the beam of light emitted by a source that is at rest in L. With respect to L this reflection is just the ordinary occurrence of light reflected by a mirror at rest. According to the principle of constancy, how-

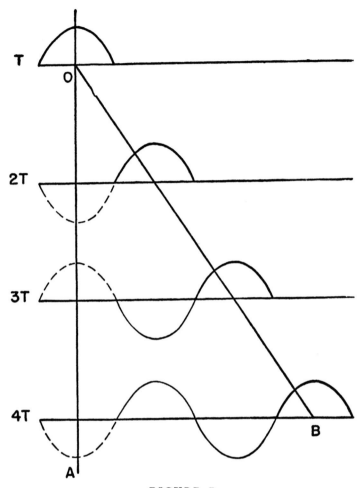

FIGURE I

This diagram represents light waves propagated in a horizontal direction through the ether. If T is a half-period of the light, the first line represents the state of the wave at the time T after the emission from the source R. The other lines represent the states of the same wave after the times 2T, 3T, and 4T respectively. If a device is placed at a fixed spot in the ether it will record the state of the wave (along the line OA) at the consequent instances of time T, 2T, 3T, and 4T. These states are represented by the interrupted lines. They indicate a vibration. But if the device of registration is moving with the speed of light in the direction of the wave propagation, it is recording the states of waves along OB. They are represented by a thick line. It is clear that *no* vibration is recorded by the moving instrument. Briefly, there is no light for a recording instrument moving with the speed of light.

ever, nothing is changed if we assume the source of light at rest in *F*. Then, however, the beam of light can never be reflected, since both the light and the mirror are traveling in the same direction with the same velocity (*c*). The light can never catch up to the mirror. Thus there would be again an influence of the speed of the laboratory on the optical phenomena within, and therefore a violation of Einstein's principle of relativity.

If we accept Einstein's two basic hypotheses, then the above considerations lead us to the conclusion: it is *not* possible for a laboratory (*L*) to move with velocity of light (*c*) with respect to the fundamental system (*F*); for if that were possible, the relativity principle could not be valid. Or, since the laboratory is a material body like any other, *no material body can move with the velocity of light* (*c*).

This conclusion may at first seem absurd. It is reasonable to think that any velocity can be attained by the continual addition of even a certain small increment of velocity. For, according to Newton's law of force, every force imparts to a body upon which it acts an additional velocity that is smaller the greater its mass is. One need only to let a force, no matter how small, act long enough on a body, and its velocity can be increased beyond any magnitude whatever. This circumstance shows the incompatibility of Einstein's principles with Newtonian mechanics; the former demand the impossibility of the velocity of light for material bodies, while the latter affords the possibility.

In Einstein's mechanics, therefore, the velocity of light in empty space plays a very special role. It is a velocity that cannot be attained or exceeded by any material body. We thus find an intimate connection created between mechanical and optical phenomena. Furthermore, owing to this circumstance, it becomes meaningful to speak of "small" or "great" velocity without further qualification. It means that the velocity is "small" or "great" in comparison with the velocity of light.

5. *Relativity of Time*

Not only did Einstein's fundamental principles give rise to results conflicting with Newtonian mechanics; they also led to drastic changes in our use of the words "space" and "time." The laws of physics contain statements about phenomena whose effects can be observed in terms of measuring rods and clocks, and much can be deduced about their behavior from Einstein's postulates.

Let us consider a situation similar to the one in the previous section. A laboratory system (L) moves with constant velocity (v, smaller than c) with respect to a fundamental system (F). There is a source of light (S) and a mirror (M) at a distance (d) from S in the laboratory (L) such that the light from S travels to M, is reflected, and returns to S, and such that the direction of the ray SM is perpendicular to the direction of the velocity v of L in respect to F. In going from the source (S) to the mirror (M) and back, the light has to travel a distance $2d$ measured by yardsticks attached to L, but by yardsticks attached to F the path is longer because the mirror (M) is moving with respect to F. Let the length of this path be $2d^*$. The ratio $\dfrac{d^*}{d}$, which will be designated by k for the sake of conciseness, is easily calculated. It requires no greater mathematical knowledge than the Pythagorean theorem, and its expression is $k = \dfrac{1}{\sqrt{1 - \dfrac{v^2}{c^2}}}$,

As v is smaller than c, k is greater than 1. k is not much greater than 1 if v is very small compared to c, but becomes very large as v approaches c.

In order to determine the dependence of (k) upon (v) we have to consider the time required for light to travel from the source S to mirror M and back to S. Some sort of a time-measuring device, such as a clock on the wall, a pocket watch on the table, a pendulum hanging from the ceiling, or an hour-glass, is needed in the laboratory (L). The time interval between the starting out of the light ray from S and its return is measured in terms of the time that the hand of a clock or watch takes to move through a certain angle, the pendulum to make a certain number of oscillations, or a certain amount of sand to flow through the hour-glass. The unit of time is a certain arbitrary angle of the clock or watch, an arbitrary number of pendulum oscillations, or an arbitrary quantity of sand.

Now, the *constancy of the velocity of light* means that the quotient of the distance traveled by the light ray divided by the time taken is equal to a constant (c), whatever the speed (v) of the source may be. The value of the distance is d if we measure it with the yardstick attached to L, and d^* if we use the yardstick attached to F. Thus if we designate the time interval for the light to go from S to M and back by t if we use the L-yard-stick and by t^* if we use the F-yardstick, we have $c = 2d/t$ and $c = 2d^*/t^*$, and hence $t^*/t = \dfrac{d^*}{d} = k$ (fig. 2). This

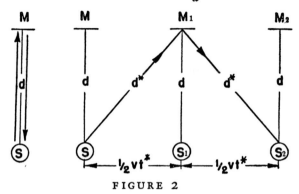

M   M M_1 M_2

FIGURE 2

The source of light S and the mirror M are moving with one and the same speed v with respect to the ether, while light itself travels at speed c. The left diagram shows a light ray emitted from S and reflected by M back to S. The line SM is the trace of the ray on a screen that participates in the motion of S and M. The time t of the relection is $t = 2d/c$, according to the principle of relativity. The right diagram shows the trace of the same light ray on a screen that is at rest in the ether and that does not participate in the motion of S and M. According to the principle of constancy we obtain $t^* = \dfrac{2d^*}{c}$. If we consider the rectangular triangle SM_1S_1, it follows from the Pythagorean theorem that $(d^*)^2 = d^2 + (vt^*/2)^2$. If we substitute the results of the principles of relativity, $d = ct/2$, and of constancy, $d^* = ct^*/2$, we obtain $\dfrac{t^*}{t} = \dfrac{1}{\sqrt{1 - \dfrac{v^2}{c^2}}} = k.$

means, however, that the result of the measurement depends on k, and consequently on v. The greater the velocity of the laboratory (L) with respect to the system (F), the greater is the angle through which the hand of the clock turns while the light travels to the mirror and back. Similarly with a pendulum and an hourglass, the greater is the number of oscillations and the quantity of sand. Therefore, by measuring this time interval, the observer in L should be able to determine the velocity (v) by observations conducted solely in his laboratory L. This, however, conflicts with Einstein's *principle of relativity*.

The contradiction arises from a traditional assumption that is based on Newton's idea of absolute time. According to Newton, all clocks, watches, hour-glasses, and any other time-measuring devices function at exactly the same rate, no matter what their velocities are. In particular, a clock in the laboratory system (L) runs at exactly the same rate as a clock firmly attached to the fundamental system (F). If this is so, t cannot differ from t^*. On the other hand, we have derived from Einstein's two hypotheses that $t^* = kt$. This means that the time t^* is different from t, and that the difference depends upon k. As k depends upon v, the rate of a time-keeper depends upon the velocity (v) of its motion. Hence if Einstein's hypotheses are accepted, the traditional assumption must be dropped, that the rate of a time-keeper is independent of its speed. In order to build up a theory of light and motion that is consistent with Einstein's hypotheses, we have to assume that the clock in the laboratory (L) runs slower than that in the fundamental system (F), the rate depending upon the speed (v) of L relative to F. Then, while the hands of the clock in F rotate through an angle (a), that of the clock in L rotate through the smaller angle a/k; while the pendulum in F makes n oscillations, that in L makes only n/k; while q ounces of sand run through the hour-glass in F, only q/k ounces run in L; hence the time interval for light to travel from S to M and back as measured by any time-measuring device attached to L will depend only on the velocity (v) of L and not on the special kind of device that we use.

Thus an entirely new property of time-keepers, which is not consistent with the traditional view, has been deduced from Einstein's two fundamental hypotheses. A moving clock, no matter what its construction, runs slower than an identical clock that is at rest. This is a physical fact that may be true or false, but there is nothing "paradoxical" about it.

Einstein even indicated a method whereby this assertion could be subjected to direct experimental verification. He pointed out that atoms could be used as natural clocks since they emit electromagnetic waves of certain definite frequencies. These frequencies of oscillation can be taken as natural time units for the atom, and frequencies of one group of atoms at rest in the laboratory can be compared with those of another group moving at a great velocity. The comparison of the frequencies can be made by means of a spectograph. The radiation of definite frequencies emitted by atoms form distinct spectral lines on photograph plates, with the position of the lines arranged according to the magnitude of the frequency. Einstein's result would be verified if the spectral lines of the moving atoms were shifted slightly to the low-frequency side as compared with the spectral lines of the stationary atoms. Actually this experiment was carried out in 1936 by H. Ives, of the Bell Telephone Laboratories, New York City, with positive result.

This effect must, of course, be distinguished from the so-called Doppler effect, which is also an alteration of the frequencies of radiation due to the motion of the atoms. The Einstein effect, however, is independent of the direction of motion of the atoms, while the Doppler effect depends critically on the direction. The shift has the greatest value if the motion of the atoms has a direction opposite to the velocity of the mirror or screen by which the light is intercepted.

There was something of a sensation when Einstein pointed out that the beat of the human heart is also a sort of clock and the rate of its beating must also be affected by its motion. Consider a person at rest in F whose heart beats at the rate of 70 per minute. If this same person moves with velocity v relative to F, then his heart will only beat $\frac{70}{k}$ times a minute. But it must be remembered that it is $\frac{70}{k}$ as measured by a clock fixed in F; if measured by a clock that travels with the person, this clock itself will move slower and the heart-beat will then be just 70. Since the same retardation likewise affects all the metabolic processes in the body, it can be said that the person moving with the system L "ages" less than a person remaining in F. Such a circumstance may sound novel, but it cannot account sufficiently for the impression that this new physical theory made upon the masses of the public. For there was an impression that all our thinking about the universe had suffered a severe shock.

In the fall of 1912 I first realized that Einstein's theory of the "relativity of time" was about to become a world sensation. At that time, in Zurich, I saw in a Viennese daily newspaper the headline: "The Minute in Danger, a Sensation of Mathematical Science." In the article a professor of physics explained to an amazed public that by means of an unprecedented mathematical trick a physicist named Einstein had succeeded in proving that under certain conditions time itself could contract or expand, that it could sometimes pass more rapidly and at other times more slowly. This idea changed our entire conception of the relation of man to the universe. Men came and went, generations passed, but the flow of time remained unchanged. Since Einstein this is all ended. The flow of time itself can be changed, and at that by a "mathematical" trick. To most people this appeared incomprehensible. Some rejoiced that anything so absurd could happen and that traditional science, which is always unpopular with some people, had suffered such a defeat. Others were vexed that something should happen which ran counter to all common sense. People were inclined to regard it as a phantasm of the mathematicians, or as an exaggeration by an author desirous of creating a sensation. At any rate, it was exciting that something of the sort could happen and that our generation was chosen to witness the overthrow of the foundations of the universe.

How did it happen that something of this kind, in part exciting and in part absurd, was ascribed to Einstein's theory? We saw above that it is actually a statement about concrete, observable events carried out with definite physical apparatuses. Why did people like to present Einstein's clear deductions about physical experiments in a semi-mystical and incomprehensive language?

The reason is that Einstein not only asserted the existence of previously unknown physical occurrences, but also proposed to describe these new phenomena in a language by which they might be expressed most simply. The usual mode of expression in physics was intended to present as simply as possible phenomena that had already been known for a long time. To Einstein this traditional language of physics proved to be too inconvenient and complicated for the presentation of the newly discovered or predicted phenomena.

In ordinary physics the duration of an event was defined by the rotation of the hands of a clock or by the number of oscillations of a pendulum. This was a clear-cut definition as long as one believed that the functioning of such a mechanism was unaffected by its motion. But if Einstein's deductions from his postulates are correct, then with moving clocks different durations of time will be obtained for the same physical event. As we saw above, the duration of the time taken for light to travel from the source (S) to a mirror (M) and back to S depends on whether the interval is measured by a clock at rest in the fundamental system (F) or by one in the laboratory (L) moving with velocity v with respect to F.

In order to express this situation most simply, Einstein proposed to speak no longer of the "duration of an event" without further qualification, but to speak of the "duration relative to a specific frame of reference." By this he meant duration measured with the aid of a clock firmly attached to this specific frame of reference. The physical situation provides no basis for selecting one of these measurements in preference to others and describing it as the "actual duration" in contrast to others that are "apparent durations." For, in accordance with the princi-

ple of relativity, the duration of a specific occurrence in a laboratory should be independent of the velocity (v) of the laboratory,
provided clocks are used that are at rest with respect to L. By no
argument, however, one can be forced to accept Einstein's proposal. One can also describe the above situation by saying: "The
true duration of an event is the duration measured by means
of the clock of a specific system of reference. Every other duration is only an illusion due to deliberate alteration in the rate
of the clock." This statement conveys exactly the same meaning
about observable facts except that a specific reference frame is
introduced, which on physical grounds is unnecessary.

Many authors have interpreted Einstein's clear and unequivo-
cal statement by the apparently profound but in reality meaningless statement: "Einstein said that sometimes time flows rapidly
and sometimes slowly." Indeed, to say that time *flows* is a figure
of speech that is only partly appropriate to the description of the
physical phenomena. To speak of "more rapid flow" is to take a
simple metaphor seriously. If one differentiates between statements about new physical occurrences and the proposal for a
new mode of expression, one can formulate what is exactly meant
by claiming the "relativity of time." It means to state: if we use
the expression "time interval with respect to a specific system of
reference," we can describe the phenomena in a simpler way
than by using the traditional expression "time interval without
specification." Einstein's *relativity of time* is a reform in *semantics,* not in metaphysics.

6. *Relativity of Other Physical Concepts*

If an investigation similar to the one on the duration
of time as measured by clocks is carried out for "intervals in
space" as measured by measuring rods, the length of a yardstick must also be affected by its motion. I shall not discuss this
point in further detail, since we have already become acquainted
with the method by which such results are obtained. I shall only
state Einstein's proposal that since moving measuring rods
change their lengths relative to resting rods, one should speak
only of "length relative to a specific system" and not of "length"
as such.

Another consequence of Einstein's basic hypotheses is that a
statement like "two events at different places occur simultaneously" is better formulated with respect to a specific system of

reference. An observer in Chicago may receive simultaneously radio signals from two points at equal distances from Chicago. He would say that they were sent out at one and the same time, but an interceptor on a moving train receiving the same signals would not receive them simultaneously if they are sent out according to conventional clocks. Einstein therefore proposed that the word "simultaneous" should likewise be introduced only in the combination "simultaneous relative to a specific system of reference." This would be again an improvement in *semantics*. "Simultaneity" without specification is an expression of little practical use.

Because of the continuity of laws, Newtonian mechanics must become invalid also for particles with velocities near to the velocity of light. Einstein soon found out that his hypotheses could be put to a very important task. They became an instrument for deriving from the laws of physics that are valid only for small velocities laws that are generally valid for all velocities. As we have learned already, it follows from Einstein's two hypotheses that Newton's law of mechanics cannot be valid for great velocities. For if it were, it would be possible, even by a small constant force, to accelerate a mass gradually until it attains the speed of light.

Einstein started from the assumption that for small velocities (i.e., much smaller than the speed of light, c) every mass moves according to Newton's laws of motion. By applying the procedure mentioned above, Einstein succeeded in deriving from the Newtonian laws the laws of motion for high speed. The chief result obtained in this way is the rather startling fact that the mass of a body is not constant; like the duration of time and the length of a measuring rod, it is dependent on its velocity. The mass increases with velocity in such a way that as the velocity becomes very great, the mass also becomes very great. A given force will produce smaller and smaller change in the actual velocity the more it approaches the velocity of light. For this reason no particle can ever actually attain the velocity of light, no matter how great a force acts on it and for how long a time.

Proceeding to the domain of electromagnetic phenomena, Einstein was again led to the conclusion that electric and magnetic field strengths are also "relative quantities." Every helpful description of electric or magnetic field strength must contain not only their magnitude but also the system with respect to which they are measured.

The necessity of this is easily seen. When an electric charge is at rest in *L*, it possesses an electric field only "relative to *L*." There is no magnetic field relative to *L* since an electric charge at rest exerts no magnetic force. However, when this same situation is described relative to *F*, the electric charge is moving with a velocity *v;* this means that there is an electric *current*. Since every electric current exerts a magnetic force, it is appropriate to say that there is magnetic field "relative to *F*." The existence of these fields is, of course, a physical fact. But their descriptions "relative to *L*" and "relative to *F*" are different.

---

7. Equivalence of Mass and Energy

From the same hypotheses Einstein was able to draw still another conclusion that at first one can hardly believe is contained in them. If an agglomeration of masses is formed or falls apart under production of kinetic energy or radiation the sum of the masses after the agglomeration or disintegration is smaller than before. The produced energy is given by $E = mc^2$, where *m* is the loss of mass. This statement may be considered as a law about the "transformation of mass into energy." In a process where there is such a transformation from mass to energy or vice versa, the energy of the system will not be conserved unless account is taken for the gain or loss due to the change in mass.

This law has proved to be of immense significance in the development of our knowledge of the interior of the atom. According to our modern conception of the atom, it consists of a massive central core with positive charge, which is called the nucleus, and around it a number of negatively charged particles, called electrons, circulating at great speed. The nucleus itself is a complex structure built up of two kinds of particles, the positively charged protons, which are the nuclei of the simplest atom, hydrogen, and neutrons that are exactly like protons except for the lack of any electric charge. The various atoms found in nature differ only by the difference in the number of protons and neutrons they possess in the nucleus, the heavier atoms containing more particles and hence being of more complex structure. As stated already, hydrogen, the lightest atom, has a nucleus that is simply a proton. The next lightest atom is helium, whose nucleus contains two protons and two neutrons. These four particles are bound very tightly together in the nucleus by certain nuclear forces. It is one of the most important problems of modern physics to investigate the strength, character, and quality of these nuclear forces which bind the atomic nuclei together.

A measure of the strength with which particles in the nucleus are packed together can be obtained by considering how much energy is necessary to pry the particles loose and separate them so they are all a large distance apart from one another. This energy is known as the binding energy of a nucleus. Now, according to Einstein's theory, this energy (E) which is produced by the formation of the nucleus must appear as loss of mass due to the agglomeration. This means that the masses of the individual protons and neutrons added together are by E/c^2 greater than the mass of the nucleus where these particles are bound together. Thus by measuring the masses of the protons and neutrons while they are free and the mass of the nucleus, it is possible to obtain the binding energy of the nucleus. Such measurements have been carried out for many of the atoms found in nature, and we are now able to classify according to how strongly the particles in the nuclei are bound together. These results have been of immense value in the planning and interpretation of recent researches on the artificial transmutation of atoms, where by bombarding various atomic nuclei with protons, neutrons, and other similar particles, new atoms have been produced.

Einstein's mass-energy relation has also for the first time in history made possible the solution of the problem of the source of the sun's energy. The sun has been radiating heat and light at the same rate as it is now doing for billions of years. If that energy had come from ordinary combustion, such as the burning of coal, the sun would have cooled off by now. The problem had the scientists completely baffled until *Einstein's equation* $E = mc^2$ appeared. The velocity of light (c) is a very large number, and with this squared, the formula states that a small quantity of mass can transform into a very large amount of energy. For this reason, by losing only an immeasurable amount of mass, the sun has been able to continue radiating for so long, and will continue to do so for billions of years to come. The actual mechanism of the transformation of mass to energy occurs in nuclear reactions that are going on in the interior of the sun. It is believed now that they ultimately boil down to the formation of helium nuclei from hydrogen. In this *"packing effect,"* as we have learned already, mass is lost and radiation emitted.

This possibility of using mass as a source of energy has aroused very optimistic hopes that methods of liberating the energy stored in the atom as mass for practical use might be found. There has also been, on the other hand, a very frightening prospect that such a process might be used to produce an explosive so devastating that a pound of it would completely annihilate everything within a radius of many miles. This foreboding was fulfilled forty years later when the first atomic bomb destroyed Hiroshima.

To Einstein, however, the main value of his result was not in the applications, no matter how numerous or important. To himself his principal achievement was to have deduced the law $E = mc^2$ from the relativity principle. It was in accord with Einstein's conception of the universe to strive continually for the discovery of simple, logical bridges between the laws of nature. The wealth of conclusions derived from his two hypotheses constitutes what has been known since as the "theory of relativity." Einstein had struck a rich well of information about nature, which would yield knowledge for many decades to come.

8. *Theory of Brownian Motion*

In the same year (1905) Einstein discovered new fundamental laws in two fields outside the theory of relativity. At the time when Einstein came to Bern, he was intensely occupied with the problem of light and motion, But he saw that the final goal could be attained only by attacking the problems from various angles. One of the paths to the goal, he realized, was to investigate the relations between light and heat, and those between heat and motion.

It had been known for some time that heat is connected with the irregular motion of molecules. The higher the temperature, the more violent is this motion. The statistical behavior of particles in such irregular motion had been investigated chiefly by the Scottish physicist James Clerk Maxwell (1831–79) and the Austrian Ludwig Boltzmann (1844–1906). It had been assumed even before, that the kinetic energy of the molecules is proportional to the absolute temperature. At the time of Maxwell and Boltzmann, however, the molecular constitution of matter was still a hypothesis which could be doubted. It enabled many different phenomena to be explained very simply, but there was as yet no very direct proof of the existence of the molecule. Furthermore, it had not yet been possible to obtain an accurate value of such a significant quantity as the number of molecules in a unit volume of matter. Estimates of this number had been made by such men as the Austrian physicist Loschmidt (1865), but they were based on involved and rather indirect methods. Einstein strongly felt the necessity of investigating this matter more thoroughly and obtaining a more direct proof of molecular motion.

It had long been known that small but microscopically visible particles, when suspended in a fluid with approximately the same density, exhibit a constant, apparently irregular zigzag motion. It had been discovered by the Scottish botanist Robert Brown for pollen dust suspended in water, and for this reason it is known as *Brownian motion*. It is not caused by any external influence jarring the vessel or by currents of water in the vessel, and the agitation increases in intensity when the temperature of the water is raised. For this reason it had been conjectured that the motion is connected with the heat motion of the molecules. According to this view, the kinetic energy of the water molecules in constant collision with the microscopic particles produces irregular forces in random directions, which give rise to the observed motions.

In 1902 Einstein had restated Boltzmann's theory of random motion in a simplified form. He now treated the Brownian motion with this method and arrived at a surprisingly simple result. He showed that the results of the kinetic theory of molecules should also hold for particles visible by microscope — for instance, that the average kinetic energy of the particles in the Brownian motion should have the same value as that for the molecules. Hence, by observing the motion of the microscopically visible particles, much valuable information could be obtained about the invisible molecules. In this way Einstein was able to derive a formula which stated that the average displacement of the particles in any direction increases as the square root of the time. He showed (1905) how one can determine the number of molecules in a unit volume by measuring the distances traveled by the visible particles.

The actual observations were later made by the French physicist Jean Perrin, who completely verified Einstein's theory. The phenomenon of Brownian motion has subsequently always been included among the best "direct" proofs of the existence of the molecule.

9. *Origin of the Quantum Theory*

To Einstein it was always clear that his theory of relativity could not claim (and, indeed, it never did claim) to solve all the mysteries of the behavior of light. The properties of light investigated by Einstein concerned only a certain group of phenomena dealing with the relation between the propagation of light and moving bodies. For all these problems light could be conceived along the lines of traditional physics as undulatory electromagnetic processes which filled space as a continuum. By the theory of relativity it was assumed that some objects can emit light of this nature, and no attempt was made to analyze the exact process by which light is emitted or to investigate whether it sufficed for a derivation of all the laws for the interaction of light with matter.

The investigations on the nature of light and its interaction with matter, however, were to lead to the rise of the "quantum theory," a revolution in physical thought even more radical than the theory of relativity. And in this field, too, Einstein's genius had a profound influence on its early development. In order to make understandable the nature of Einstein's contributions, I shall describe briefly the situation prior to his researches.

The simplest way of producing light is by heating a solid body. As the temperature rises, it begins to glow from a dull cherry red to a brighter orange, and then to blinding white light. The reason for this is that visible light consists of radiations of different frequencies ranging from red at the low end through the colors of the spectrum up to violet at the high end. The quality of light emitted by a solid body depends solely on its temperature; at low temperatures the low-frequency waves predominate and hence it looks red; at higher temperatures the shorter wave lengths appear and mingle with the red to give the white color.

Attempts to explain this change in quality of light with temperature on the basis of nineteenth-century physics had ended in failure, and this was one of the most important problems facing physicists at the beginning of the twentieth century. At that time the emission of light was thought to be produced by the oscillations of charged particles (electrons), the frequency of light emitted being equal to the frequency of the vibration. According to Boltzmann's statistical law, already mentioned, the average energy of oscillation of an electron should be exactly equal to the average kinetic energy of gas molecules, and hence simply proportional to the absolute temperature. But this led to the conclusion that the energy of vibrations is independent of the frequency of oscillation, and hence light of different frequencies will be emitted with the same energy. This conclusion obviously was contradicted by the observations on light emitted by heated bodies. In particular, we know that light of very short wave lengths is not emitted to any great extent by hot bodies. As the temperature increases, rays of increasingly higher frequencies appear, but yet at a given temperature there is no perceptible radiation above a certain definite frequency. Consequently it appeared that somehow it must be difficult to emit light of very high frequencies.

Since all arguments based on the mechanistic theory of matter and electricity led to results conflicting with experience, the German physicist Max Planck in the year 1900 introduced a new assumption into the theory of light emission. At first it appeared to be rather inconsequential, but in the course of time it has led to results of an increasingly revolutionary character. The turn in physics coincided exactly with the turn of the century. I shall sketch Planck's idea in a somewhat simplified and perhaps superficial form.

According to Boltzmann's statistical law, the average energy of oscillation of an electron in a body is equal to the average kinetic energy of the molecules. The actual energies of the individual atoms or molecules can, of course, have very different values; the statistical law only relates the *average* energy with the temperature. Boltzmann, however, had been able to derive a second result which determined the distribution of the energy of the particles around the average value. It stated that the number of particles with a certain energy depends on the percentage by which this energy differs from the average value. The greater a deviation, the less frequent will be its occurrence.

As Planck realized, the experimental results indicated that the oscillating electrons in a body cannot emit radiation with an arbitrary frequency. The lack of high-frequency radiation shows that the mechanism of radiation must be such that it is somehow difficult to emit light of high frequency. Since no explanation of such a mechanism existed at that time, Planck was led to make the new assumption that, for some reason as yet unknown, the energy of oscillation of the atoms cannot have just any value, but can only have values that are integral multiples of a certain minimum value. Thus, if this value is called ϵ, then the energy of the oscillations can only have the discrete values *0*, ϵ, 2ϵ . . . or *n*ϵ, whose *n* is zero or an integer. Consequently the radiation emitted or absorbed must take place in portions of amount ϵ. Smaller amounts cannot be radiated or absorbed since the oscillation cannot change its energy by less than this amount. Planck then showed that if one wants to account for the well-known fact that a shift to higher temperatures means a shift to higher frequencies, one has to take values for ϵ that vary for different values of the frequencies of the oscillations, and in fact ϵ has to be proportional to the frequency.

Thus he put ϵ = *hv*, where *v* is the frequency and *h* is the constant of proportionality, which has since then been called Planck's constant and has been found to be one of the most fundamental constants in nature. With this assumption Planck was immediately able to derive results in the theory of radiation that agreed with observations and thus removed the difficulties that had confronted the physicists in this field.

---

10. *Theory of the Photon*

Planck thought he was only making a minor adjustment in the laws of physics in formulating his hypothesis, but Einstein realized that if this idea was developed consistently it would lead to a rupture of the framework of nineteenth-century physics so serious that a fundamental reconstruction would be necessary. For if the electron can oscillate only with certain discrete values of energy, it contradicts Newton's laws of motion, laws which had been the bases for the whole structure of mechanistic physics.

Planck's hypothesis dealt only with the mechanism of radiation and absorption of light and stated that these processes could take place only in definite amounts. He said nothing about the nature of light itself while it is propagated between the point of radiation and that of absorption. Einstein set out to investigate whether the energy transmitted by light retained this discrete character during its propagation or not. He once expressed this dilemma by the following comparison: "Even though beer is always sold in pint bottles, it does not follow that beer consists of indivisible pint portions."

Retaining the analogy, if we wish to investigate whether the beer in a barrel actually consists of definite portions or not, and if so whether this portion is a pint, two pints, or ten pints, we can proceed as follows: We take a number of containers, say ten to be definite, and pour the beer from the barrel at random into these containers. We measure the amount in each container and then pour back the beer into the barrel. We repeat this process a number of times. If the beer does not come in portions, the average value of the beer poured into each container will be the same. If it consists of pint portions, there will be variations in the average values. For two pints the variations will be greater, and for ten they will be still greater. Thus by observing the distribution of beer among the ten vessels, we can tell whether the barrel of beer consists of portions and what size they are. We can realize it easily by imagining the extreme case that the whole content of the barrel is one portion.

The situation is similar in the case of radiation enclosed in a box. We can imagine this box to be divided into a number of cells of equal volume and consider the distribution of the energy of radiation in these cells. If the portions of radiation are large, the variations of energy among the cells will be large, and if they are small, these variations will be small. From the empirical law of distribution it follows that the variations in the violet light are greater than in the red light. Einstein drew the conclusions that violet light consists of a few large portions, while red light consists of many small ones. Exact calculations showed that the magnitude of the portions must be $h\nu$. Thus Einstein found that not only did emission and absorption of radiation take place in discrete amounts, but light itself must consist of definite portions. The name "photon" has since been given to the quantum of radiation.

To this conclusion, which Einstein derived theoretically, he was able to point out an experimental verification. It had been known for some time that when light shines on certain metals, electrons are given off. Electrons are fundamental particles in physics which carry a negative electric charge and constitute the outer portion of the atom. In 1902 the German physicist Philipp Lenard discovered a very startling result of this emission of electrons. He found that the intensity of light falling on the metal had no effect on the energy with which the electrons are ejected from the metal, but that this energy depends only on the color or frequency of the light. No matter how far the source of light is moved away from the metal, the electrons are still ejected with the same velocity, though of course the number ejected is smaller. But when violet light is used instead of red, the velocity of the electrons is much greater.

According to Einstein's view, the explanation is quite simple. No matter what distance light of a certain color has traveled from the source, it still consists of the same portions of energy, the only difference being that, farther away from the source, the individual portions are spread out thinner. The ejection of an electron occurs when a whole quantum of radiation is absorbed by a single electron, which then comes off with the energy of the photon. Thus the distance between the source and the metal has no effect on the energy of the single electron emitted. Furthermore, the difference between violet and red light is that a different amount of energy is possessed by the photon. Hence an electron that absorbs a violet photon naturally comes off with higher velocity than one that absorbs a red photon.

To form another analogy, let us consider the bombardment of a fortification by machine guns and by heavy artillery. Even if the total weight of projectiles fired is the same in both cases, the effects produced are of a very different character. The machine-gun bullets make a very large number of small dents, while the artillery shells make a few big holes. Moreover, the average intensity of gunfire has very little effect on the magnitude of the holes, but only on their number.

With his hypothesis on the discontinuous nature of light Einstein threw doubt on the entire conception of a continuous field of force. If light consists of photons, the electric and magnetic fields cannot fill all space continuously, and the whole of the electromagnetic theory of light based on this concept has to be re-examined. The discontinuous structure is apparently inconsistent, however, with some observed phenomena, in particular with interference and diffraction of light, which are explained so well by the theory of continuous waves. Einstein, who was well aware of this difficulty, looked upon his assumption only as a provisional hypothesis, without any lasting value. He therefore entitled the paper in which he presented his discovery: "On a Heuristic Point of View Regarding the Production and Transformation of Light."

It is interesting to note that Einstein's new quantum theory of light was based upon the research of two German physicists who were later to play important roles in his life. Max Planck was first to advocate the significance of Einstein's theory of relativity, and Philipp Lenard was to oppose it most vehemently on philosophical, political, and racial grounds.

IV

EINSTEIN AT PRAGUE

---

1. *Professor at the University of Zurich*

The researches whose results Einstein published at Bern in 1905 were so unusual that to the physicists of the Swiss universities they seemed incompatible with the assigned work of a minor official of the patent office. Attempts were soon made to bring Einstein to teach at the University of Zurich. At this time Professor Kleiner was the leading personality in physics there. He was a man who realized that Einstein's papers revealed an unusual talent, but who did not really understand them. He felt it his duty to do the best for his university and endeavored to appoint Einstein professor at Zurich.

According to the regulations in force at Zurich as well as at other Germanic universities, no one could be appointed professor at a university unless he had previously been a *Privatdozent*. This is a position for which there is no analogue in the universities of western Europe and America. A young man with scientific achievements may apply for permission to teach at a university. He has no obligations and can lecture as much or as little as he desires, but receives no remuneration except the usually very small fees paid by students who attend his lectures. Since for this reason the number of *Privatdozenten* does not have to be restricted, this system has the advantage that every young scientist has an opportunity to show his teaching abilities and the universities have a large number of candidates to choose from in appointing their professors. The disadvantage, of course, is that in practice only persons with private means or another position which supports them can enter this career. With his position at the patent office, Einstein was in the latter situation.

Professor Kleiner advised him to become a *Privatdozent* at the university of Bern, so that after a short while he could then be eligible for a professorship at Zurich. Although he did not like the idea of giving regular lectures, Einstein followed the advice. Consequently his lectures were not very well prepared, and since the students were not obliged to attend them, only a few friends came. Furthermore, Einstein was then in the midst of a veritable maelstrom of new discoveries and it was difficult to arrange his material in a way appropriate to the capacities of the average student. Professor Kleiner once came to Bern to hear Einstein lecture and afterward remarked to him that such lectures did not seem on a level fitted for the students. Einstein answered: "I don't demand to be appointed professor at Zurich."

At that time the professorship of theoretical physics at the University of Zurich became vacant, but the board of education of the canton of Zurich, which was in charge of the university, had its own plans for this position. The majority of the board of education belonged to the Social Democratic Party, and they had in Zurich a party comrade who appeared to be a suitable candidate, from both the political and the scientific viewpoint. This man was Friedrich Adler, Einstein's former fellow student at the Zurich Polytechnic, who was then a *Privatdozent* at the University of Zurich. As the son of the leader of the Austrian Social Democrats, he was held in high esteem by the party members in Zurich. Friedrich Adler was a man imbued with a fanatical love of truth and was interested in physics chiefly because of its philosophical aspects. He was in every respect a man who would not shrink from uttering what he regarded as the truth even if it was to his own disadvantage. Learning that it was possible to obtain Einstein for the university, he told the board of education: "If it is possible to obtain a man like Einstein for our university, it would be absurd to appoint me. I must quite frankly say that my ability as a research physicist does not bear even the slightest comparison to Einstein's. Such an opportunity to obtain a man who can benefit us so much by raising the general level of the university should not be lost because of political sympathies."

So in 1909, despite the political leaning of the board of education and the leading professor's disapproval of his mode of lecturing, Einstein was appointed professor "extraordinary" at the University of Zurich.

The call to Zurich gave Einstein for the first time a position with a certain public prestige. Most *Privatdozenten* feel that they have become important persons when they attain professorial rank, for then they can lord it over the *Dozenten* instead of being passive objects to be dealt with by the university administration. For Einstein this was naturally no cause for satisfaction. He had not suffered in any way while a *Privatdozent*, and he did not have any desire to dominate others. Besides, he had not been anxious enough for the position to derive any great pleasure from its attainment.

From the financial point of view, the position of a professor "extraordinary" was not very lucrative. His income was no larger than it had been at the patent office and, moreover, he could no longer lead an inexpensive and pleasant bohemian life now that he had acquired a certain social status in the city. Though he kept expenses at a minimum, he had to spend money for things from which he derived no pleasure, but which were required by his social position. In order to improve the financial situation, his wife took in students to board. He once said jokingly: "In my relativity theory I set up a clock at every point in space, but in reality I find it difficult to provide even one clock in my room."

Einstein loved the city of Zurich, which had become his home. His wife also felt more at home here than anywhere else. Collaboration with students and colleagues, which was now possible, was a great stimulus to Einstein. Administrative duties and regular teaching, however, had few attractions and in certain respects many difficulties. This was due not only to the constraint a person of such great creative ability finds himself under when required to expend his efforts on tasks that do not appear important, but also to Einstein's paradoxical relation to society, arising from his personality.

The immediate impression that Einstein made on his environment was a conflicting one. He behaved in the same way to everybody. The tone with which he talked to the leading officials of the university was the same as that with which he spoke to his grocer or to the scrubwoman in the laboratory. As a result of his great scientific discoveries, Einstein had already acquired a profound inner feeling of security. The pressure that had often burdened his youth was gone. He now saw himself in the midst of the work to which he was going to devote his life and to which he felt himself equal. Alongside this work the problems of daily life did not appear very important. Actually he found it very difficult to take them seriously. His attitude in intercourse with other people, consequently, was on the whole one of amusement. He saw everyday matters in a somewhat comical light, and something of this attitude manifested itself in every word he spoke; his sense of humor was readily apparent. When someone said something funny, whether intentionally or not, Einstein's response was very animated. The laughter that welled up from the very depth of his being was one of his characteristics that immediately attracted one's attention. To those about him his laughter was a source of joy and added to their vitality. Yet sometimes one felt that it contained an element of criticism, which was unpleasant for some. Persons who occupied an important social position frequently had no desire to belong to a world whose ridiculousness in comparison to the greater problems of nature was reflected in this laughter. But people of lesser rank were always pleased by Einstein's personality.

Einstein's conversation was often a combination of inoffensive jokes and penetrating ridicule, so that some people could not decide whether to laugh or to feel hurt. Often the joke was that he presented complicated relationships as they might appear to an intelligent child. Such an attitude often appeared to be an incisive criticism and sometimes even created an impression of cynicism. Thus the impression Einstein made on his environment vacillated between the two poles of childish cheerfulness and cynicism. Between these two poles lay the impression of a very entertaining and vital person whose company left one feeling richer for the experience. A second gamut of impression varied from that of a person who sympathized deeply and passionately with the fate of every stranger, to that of a person who, upon closer contact, immediately withdrew into his shell.

2. *Appointment to Prague*

In the fall of 1910 there occurred a vacancy in the chair of theoretical physics at the German University in Prague. Such appointments were made at the recommendation of the faculty by the Emperor of Austria, who exercised his right through the Ministry of Education. The decisive man in the selection of the candidate was the physicist Anton Lampa, a man of very progressive tendencies as far as education was concerned. All his life he fought for the introduction of modern pedagogical methods, for the freedom of teaching from reactionary influences, and for the extension of scientific and artistic education to the largest possible number of the population. There was a considerable gap between his high aspirations and his scientific capacities, however, and as a result he was animated by an ambition he could not satisfy. Since he was a man of high ethical ideals, he consciously sought to suppress this ambition, but the result was that it played an even greater role in his subconscious life. His philosophical *Weltanschauung* was for the most part determined by the positivistic philosophy of the physicist Ernst Mach, whose student he had been. It was Lampa's life goal to propagate Mach's views and to win adherents for them.

When the question of filling the chair of theoretical physics came up, Lampa thought that here was an opportunity to appoint someone who would teach physics in the spirit of Mach. In addition, it had always been his dream to enter the realm of the extraordinary and of the genius, and he wanted an outstanding scientist, not an average professor. Even though he realized that he himself was not so gifted, he was just enough to accept the presence of an outstanding man.

Lampa had in mind two physicists who he thought would teach in the spirit of Mach and were acknowledged to have extraordinary capacities. The first was Gustav Jaumann, a professor at the Technical Institute in Brno, and the second was Einstein. Jaumann followed Mach in some peculiarities, chief among which was his aversion to the introduction of atoms and molecules in physics. Even when the atomic constitution of matter had been generally accepted as giving the best and simplest presentation of physical phenomena, Jaumann retained Mach's predilection and tried to build up a theory of continuously distributed matter. Since he had a great natural talent and imagination, he considered himself a neglected genius and developed an excessive vanity and sensitivity. Einstein, on the other hand, was influenced more by the spirit than by the letter of Mach's teachings. We have already seen in his work on Brownian motion that Einstein did not follow Mach's rejection of the atom.

Since the regulations provided that the names of the proposed candidates be listed on the basis of their achievements, Einstein, whose writings in the years from 1905 to 1910 had already made a strong impression on the scientific world, was placed first and Jaumann second. Nevertheless, the Ministry of Education first offered the position to Jaumann. The Austrian government did not like to appoint foreigners and preferred Austrians. But the ministry had not taken Jaumann's vanity and touchiness into account. He said: "If Einstein has been proposed as first choice because of the belief that he has greater achievements to his credit, then I will have nothing to do with a university that chases after modernity and does not appreciate true merit." Upon Jaumann's rejection of the offer, the government overcame its aversion to foreigners and offered the position to Einstein. He had some qualms about going to a foreign country, and his wife did not want to leave Zurich, but eventually he accepted it. One deciding factor was the circumstance that for the first time in his life he was to have a full professorship with adequate salary.

There was one peculiar difficulty to be overcome, however, in taking up the position. The octogenarian Emperor Franz Josef was of the opinion that only a man who belonged to a recognized church should be a teacher at a university, and he refused to confirm the appointment of anyone who did not conform to this rule. Einstein's friends at the university who had proposed his appointment informed him of this circumstance. Since leaving the gymnasium in Munich, Einstein had not been an official member of any religious community, but in order to avoid this difficulty, he indicated that he was an adherent of the Jewish religion, to which he had belonged as a child. He did not go through any formal ceremony, but in the questionnaire that he had to fill out he simply wrote his religion was "Mosaic," as the Jewish creed was then called in Austria.

When Einstein arrived in Prague he looked more like an Italian virtuoso than a German professor, and he had, moreover, a Slav wife. He was certainly unlike the average professor at the German University. Since he had been preceded by the reputation of being not an ordinary physicist but an extraordinary genius, everyone was curious to meet him.

In Prague it was the custom for a newly arrived member of the faculty to pay a call on all his colleagues. In his good-natured way Einstein was ready to accept the advice of his friends and make the necessary calls, which numbered some forty. He also decided to take advantage of this opportunity to see various sections of the romantic old city of Prague, and so he began to make his visits according to the location of the houses. All who made his personal acquaintance were immediately pleased by his naturalness, his hearty laughter, and the friendly and at the same time dreamy look in his eyes. But Einstein soon began to find the calls rather a nuisance. He felt that it was a waste of time to carry on conversations about trivialities and suddenly he stopped his visits. The professors upon whom he had not called were puzzled and offended at this neglect. Some people began to regard him as either proud or capricious, when the true explanation was that these colleagues lived in sections of the city that did not interest Einstein, or their names were too far back in the faculty directory.

This aversion to all formality and ceremonial was a very important trait in Einstein's character. It was particularly marked for ceremonies that were in any way depressing. Thus Einstein had an intense aversion to attending funerals, and on one occasion when he was in a funeral procession he remarked to his assistant, walking at his side: "Attending funerals is something one does to please the people around us. In itself it is meaningless. It seems to me not unlike the zeal we polish our shoes with every day just so that no one will say we are wearing dirty shoes." Throughout his life Einstein had maintained this attitude of revolt against the customs of bourgeois life.

3. Colleagues at Prague

The University of Prague is the oldest university in central Europe. During the second half of the nineteenth century there had been German and Czech professors lecturing in their respective languages, but with political quarrels creating more and more difficulties, the Austrian government had in 1888 decided to divide the university into two parts, thus creating a German and a Czech university. It is perhaps an interesting historical accident that the first rector of the German University, where Einstein was appointed, had been Ernst Mach.

At the time of Einstein's arrival the two universities were completely separated and there were no relations between the professors of the two institutions. Even professors of the same subject had no personal contact and it frequently happened that two chemistry professors from Prague would meet for the first time at an international congress in Chicago. There was already a group among the Germans who propagated the idea of the "master race" and frowned upon any intercourse with "inferior races." The majority of the German professors had too little interest in politics or were too timid to oppose the powerful will of this group by entering into contact with the Czechs.

Nevertheless, the general attitude of superiority and hostility against the Czechs was quite evident in the conversations among the German professors and their families. Comical stories were told of how Czechs behaved in society for which, in the Germans' opinion, they were not suited. The situation may be described by the following instances:

During a population census undertaken by the government, a professor of political science sent a circular letter to the mem-

Einstein at the time of his most intense scientific work

*Einstein, Paul Ehrenfest, Paul Langevin, Kammerling-Onnes, and
Pierre Weiss at Ehrenfest's home, Leyden, the Netherlands*

bers of the university faculty urging them to list all their serv-
ants as German even if they were Czech. He reasoned as follows:
Servants should only speak to their masters; since the latter are
German, the language of all the servants must be German.

Another professor, while walking with a colleague one day,
saw a house sign that seemed about to fall down on the sidewalk.
"It doesn't matter much," he said, "since it is extremely probable
that when it falls it will strike a Czech."

One of the remarkable and frequently comical aspects of this
hostility was that there was not even the slightest difference between the Germans and the Czechs in Prague so far as race and
origin were concerned. The question of which nationality one
belonged to was often a question of personal taste and which
offered opportunities for earning a living.

Anton Lampa, Einstein's closest colleague, was the son of a
Czech janitor. But, as frequently happened among the Czechs,
the son had worked his way up, driven by his ambition and a
great desire for knowledge and learning. Though his father was
a Czech, he worked in a building belonging to Germans, so
young Lampa attended German schools. He spoke Czech and
German with equal facility, and upon graduating from the gymnasium he was faced with the problem of deciding whether to
attend the German or the Czech university. He chose the former
and later became a student of Ernst Mach. Yet despite his past
Lampa was just as hostile to the Czechs as the other Germans.
He was one of those who, for instance, refused to buy a postcard if the word "postcard" was printed on it in both languages,
and demanded a card having only the German word on it. If
the post-office clerk was a Czech, he would frequently say that
such cards had all been sold out. The professor would then argue
that it was the clerk's duty to keep cards with purely German
text, and so a quarrel would begin.

Under these circumstances it was difficult even for a German
who disapproved of this hostile attitude to come into contact
with the Czechs. The latter were very suspicious and sensitive
and felt insulted by every thoughtless word. They suspected
everyone of wishing to humiliate and disparage them, and as a
result it was not easy for a well-meaning German to maintain
friendly relations with the Cechs. It is not surprising, therefore,
that Einstein hardly came in contact with them. He disapproved
the standpoint of his colleagues and did not join in their disparaging anecdotes, but he did not become intimately acquainted
with any Czechs. But Czech students did attend his lectures and

carry on scientific research under his direction, in itself a rare occurrence at the German University.

Among his closest colleagues Einstein was attracted most strongly by a mathematician named Georg Pick. He was some twenty years older than Einstein and was an extraordinary personality, both as a man and as a scientist. Pick was above all a creative mind in mathematical research. In very concise papers he published many precisely formulated ideas, which were later developed by others as independent branches of mathematics. Nevertheless, he never received much of the scientific recognition he deserved, since he was of Jewish ancestry and had rather an uncompromising nature. He held firmly to what he considered was right and did not make concessions of any kind. After his retirement at an age of over 80, he died in a Nazi extermination camp.

As a young man Pick had been an assistant of Ernst Mach's when Mach was professor of experimental physics at Prague. Einstein liked to hear Pick reminisce about Mach, and Pick was particularly fond of repeating statements by Mach that could be interpreted as anticipating Einstein's theories. Pick was also a good violinist, and through him Einstein became acquainted with a group of music-lovers and was urged to participate in chamber music. After that, Einstein had his regular quartet evenings.

Einstein and Pick met almost daily and they discussed many problems together. In the course of long walks Einstein confided to Pick the mathematical difficulties that confronted him in his attempts to generalize his theory of relativity. Already at that time Pick made the suggestion that the appropriate mathematical instrument for the further development of Einstein's idea was the "absolute differential calculus" of the Italian mathematicians Ricci and Levi-Civita.

Einstein's immediate assistant at this time was a young man named Nohel. He was the son of a small Jewish farmer in a Bohemian village, and as a boy he had walked behind the plow. He had the quiet poise of a peasant rather than the nervous personality so often found among the Jews. He told Einstein a good deal about the condition of the Jews in Bohemia, and their conversations began to arouse Einstein's interest in the relation between the Jews and the world around them. Nohel told him about the Jewish peasants and tradesmen who in their daily activities used the Czech language. On the Sabbath, however, they spoke only German. For them this language, so close to

Yiddish, was a substitute for Hebrew, which had long since been given up as the language of daily life.

Another colleague with whom Einstein became quite intimate was Moritz Winternitz, a professor of Sanskrit. He had five children to whom Einstein became greatly devoted, and he once remarked: "I am interested to see how a number of such commodities produced by the same factory will behave." Professor Winternitz had a sister-in-law who very often accompanied Einstein at the piano when he played the violin. She was an elderly maiden lady whose life had been spent in giving piano lessons and who had thus acquired a somewhat dictatorial manner. She used to speak to Einstein as if she were addressing a pupil. Einstein often remarked: "She is very strict with me," or "She is like an army sergeant."

When Einstein was to leave Prague, he had to promise her that he would recommend as his successor as professor of theoretical physics only someone who could also replace him as her violin partner. When I went to Prague to replace Einstein and was introduced to her, she immediately insisted that I keep this promise by playing the violin. To my regret, I had to tell her I had never in my whole life had a violin in my hands. "So," she replied, "Einstein has disappointed me."

---

4. *The Jews in Prague*

The appointment as professor at Prague led Einstein to become a member of the Jewish religious community. Even though this relation was only formal and the contact was only a very loose one at that time, it was in this period of his life that perhaps for the first time since his childhood he came aware of the problems of the Jewish community.

The position of the Jews in Prague was a peculiar one in many respects. More than half of the German-speaking inhabitants in Prague were Jewish, so that their part among the Germans, who comprised only about five per cent of the total population, was extraordinarily important. Since the cultural life of the Germans was almost completely detached from that of the Czech majority, with separate German theaters, concerts, lectures, balls, and so on, it was not surprising that all these organizations and affairs were dependent on Jewish patronage. Consequently, for the great masses of the Czech people, a Jew and a German were approximately the same. At the time when Einstein came to Prague, the World War I was just in the making and the Czechs felt that they were being driven into a war by the government against their own interests but in the interests of the hated Germans. They looked upon every German and Jew as a representative of a hostile power who had settled in their city to act as a watchman and informer against the Czech enemies of Austria. There is no doubt that there were some Jews, who, aping other Germans, somehow adapted themselves to this role of being policemen and instruments of oppression. But the core of the Jewish population was disgusted.

On the other hand, the relation of the Jews to the other Germans had already begun to assume a problematical character. Formerly the German minority in Prague had befriended the Jews as allies against the upward-striving Czechs, but these good relations were breaking down at the time when Einstein was in Prague. When the racial theories and tendencies that later came to be known there as Nazi creed were still almost unknown in Germany itself, they had already become an important influence among the Sudeten Germans. Hence a somewhat paradoxical situation existed for the Germans in Prague. They tried to live on good terms with the Jews so as to have an ally against the Czechs. But they also wanted to be regarded as thoroughly German by the Sudeten Germans, and therefore manifested hostility against the Jews. This peculiar situation was characterized outwardly by the fact that the Jews and their worst enemies met in the same cafés and had a common social circle.

At this time in Prague there was already a Jewish group who wanted to develop an independent intellectual life among the Jews. They disliked seeing the Jews taking sides in the struggle between Germans and Czech nationalists. This group was strongly influenced by the semi-mystical ideas of the Jewish philosopher Martin Buber. They were Zionists, but at that period they paid little attention to practical politics and concerned themselves mainly with art, literature, and philosophy. Einstein was introduced to this group, met Franz Kafka, and became particularly friendly with Hugo Bergmann and Max Brod.

Hugo Bergmann was then an official in the university library. He was a blond young man with a gentle, intelligent, and yet energetic personality. He was the center of a youthful group in Prague that attempted to create a Jewish cultural life not based on orthodox Judaism, which approached the non-Jewish world with sympathetic understanding, not aversion or blind imitation. Bergmann based his theories not only on Jewish authors but also on German philosophers such as Fichte, who preached the cultivation of the national spirit.

Even such an intelligent and ardent Zionist as Bergmann, however, could not interest Einstein in Zionism for the time being. He was still too much concerned with cosmic problems, and the problems of nationality and of the relation of the Jews with the rest of the world appeared to him only as matters of petty significance. For him these tensions were only expressions of human stupidity, a quality that on the whole is natural to man and cannot be eradicated. He did not realize then that these troubles would take on later cosmic dimensions.

At this time Max Brod was a young writer of multifarious interests and talents. He was also very much interested in historical and philosophical problems, and in his novels he described the life of the Czech and other inhabitants of Prague and Bohemia. His novels were characterized by clear, rather rationalistic analyses of psychological processes.

In one of his novels, *The Redemption of Tycho Brahe,* he described the last years of the great Danish astronomer Tycho Brahe, which were spent in Prague. The chief theme of the novel is the antithesis of the character of Tycho and of the young astronomer Kepler, whom the former had invited to work with him so as to have a collaborator who would add his young unprejudiced creative ideas to Tycho's great experience and powers of observation. It was often asserted in Prague that in his portrayal of Kepler, Brod was greatly influenced by the impression that Einstein's personality had made on him. Whether Brod did this consciously or unconsciously, it is certain that the figure of Kepler is so vividly portrayed that readers of the book who knew Einstein well recognized him as Kepler. When the famous German chemist W. Nernst read this novel, he said to Einstein: "You are this man Kepler."

5. *Einstein's Personality Portrayed in a Novel*

It therefore seems appropriate to quote several passages where Brod characterizes his Kepler and in which we may perhaps find certain aspects of Einstein's personality. The words of a poet may be more impressive than the description of a scientist.

Kepler's calm, quiet nature sometimes aroused a feeling of uneasiness in the passionate Tycho. Brod describes Tycho's feelings toward Kepler in a way that is probably equally true of the attitudes of Einstein's scientific colleagues toward him:

"Thus the storm raged in Tycho's spirit. He took the greatest pains to keep his feelings for Kepler free from alloy. . . . In actual fact he really did not envy Kepler his success. At the very most, the self-evident and in all respects becoming and worthy manner in which Kepler had achieved renown sometimes excited in him an emotion bordering upon envy. But in general Kepler now inspired him with a feeling of awe. The tranquillity with which he applied himself to his labors and entirely ignored the warblings of flatterers was to Tycho almost superhuman. There was something incomprehensible in its absence of emotion, like a breath from a distant region of ice. . . . He recalled that popular ballad in which a *Landsknecht* had sold his heart to the Devil and had received in exchange a bullet-proof coat of mail. Of such sort was Kepler. He had no heart and therefore had nothing to fear from the world. He was not capable of emotion or of love. And for that reason he was naturally also secure against the aberrations of feelings. 'But I must love and err,' groaned Tycho. 'I must be flung hither and thither in this hell, beholding him floating above, pure and happy, upon cool clouds of limpid blue. A spotless angel! But is he really? Is he not rather atrocious in his lack of sympathy?' "

This appearance of pure happiness, however, which the superficial observer was frequently inclined to ascribe to Einstein likewise, is certainly only an illusion. Tycho, who, as is well known, was the inventor of a cosmic system that represented a kind of compromise between the old Ptolemaic and the new Copernican system, was very curious to hear Kepler's opinion of this system. He always suspected that in his heart Kepler favored Copernicus and his radically new theory. Kepler, however, avoided the expression of any definite opinion on this subject before Tycho. He discussed only concrete astronomical problems with him, no general theories. Tycho felt that this was an evasion and urged him to talk about it. Finally Kepler answered him:

" 'I have little to say. . . . I am still undecided. I can't come to a decision. Besides, I don't think that our technical resources and experience are yet sufficiently advanced to enable us to give a definite answer to this question.' "

There was a pause, during which Kepler sat completely self-absorbed, with a blissful smile on his countenance. But Tycho was already somewhat irritated and interrupted him:

" 'And does this satisfy you, Kepler, this state of affairs? I mean this uncertainty regarding the most essential points of our art. Doesn't the lack of decision sometimes take your breath away? Doesn't impatience deprive you of all your happiness?'

" 'I am not happy,' Kepler answered simply. 'I have never been happy.'

" 'You not happy?' Tycho stared at him with wide-open eyes. 'You — not — what do you lack, then? What more do you want? What would you have in addition to that already bestowed on you? — Oh, fie, how immodest you must be if you don't reckon yourself happy, you who are the happiest of all men! Yes, must I, then, tell it to you for the first time? Don't you feel that you — now I will put it in one word, that you are on the right way, on the only right way? . . . No, now I don't mean the outward success, the applause surrounding you, which has been accorded you. But inwardly, inwardly, my Kepler — must I really say it to you? — inwardly, in the heart of our science, you are on the right path, the path blessed by God; and that is the noblest, happiest fate that a mortal can encounter.'

" 'No, I am not happy, and I have never been happy,' Kepler repeated, with a dull obstinacy. Then he added quite gently: 'And I don't wish to be happy.'

"Tycho was at his wits' end. . . . But even while he labored to represent Kepler to himself as a cunning, calculating man, an intriguer, it was fully clear to him that this in no way tallied with the facts, that Kepler was the very opposite of an intriguer; he never pursued a definite aim and in fact transacted all affairs lying outside the bounds of his science in a sort of dream. Why, he did not even realize that he was happy. So far did his mental confusion go that he did not even observe that. . . . He was not responsible for anything that he did. . . . With all his happiness, which another man would have had to purchase at the expense of unending suffering on the part of his conscience, Kepler was pure and without guilt; and this absence of guilt was the crown of his happiness; and this happiness — thus the circle closed — did not for a moment weigh upon him, for he was not even conscious of it. . . . He really had no inkling of his good fortune. There he sat at the table opposite Tycho, and while Tycho was tossed hither and thither by his thoughts, he sat with upright, with somewhat rigid torso, in the attitude of one whose gaze is fixed upon the distance, sat in complete calm and composure, observing nothing of Tycho's disquiet and — as usual continued calculating."

On another occasion Kepler and Tycho again discussed the arguments favoring or opposing either the Copernican or the Tychonian system, and this time they paid more attention than before to the observable facts and the logical conclusions that could be drawn upon for such proof. Brod describes the attitudes of the two men as follows:

Tycho "began to despair, finding no sign of decision on either side. Kepler, on the other hand, seemed to drink in a copious draught of pleasure and strength from this very uncertainty. The more obscure and the more difficult the decision, the more did he find himself in the humor for jesting, this man who was ordinarily so dry. When confronted by 'Nature,' this riddle of the Sphinx, his whole being expanded, he seized without difficulty upon the object, jovially assailing it upon every side, as it were, and firmly rooted himself in it. His voice even took on an unfamiliar, joyously consequential bassnote when he cried in reply to a caustic remark of Tycho's: 'Well, perhaps the laws of nature agree only fortuitously.'"

Another discussion develops between Tycho and Kepler over the question whether scientists in espousing a hypothesis must consider the beliefs and opinions of rulers and rich men.

"Tycho raised himself, breathing heavily. 'Now at least the system of Copernicus remains unproved, and as it runs counter to the Bible and as I may not needlessly affront the Catholic Majesty of my Emperor, I have no reason for espousing it.'

"'That is going too far,' observed Kepler, still smiling. 'Catholic or not, the hypothesis alone is being considered here, not the Emperor's favor.' . . .

"Tycho answered hotly, feeling that a fundamental principle of his life was being assailed: 'But without the favor of princes and of the rich we could construct no expensive apparatus, and truth would remain uninvestigated. . . . Thus the princes help us and the truth; so it is for us in our turn to respect them and to defer to their pleasure.'

"'It is just this that I contest,' cried Kepler excitedly; 'we must defer to truth alone and to no one else. . . .'

"'Why to no one else? . . . When I have already put it before you that one can serve the truth only if one serves princes. It is quite true that it is more comfortable and simpler to follow your practice, my dear Kepler. You pay regard to nothing in going your own holy way, turning neither to right nor to left. But does it seem less holy to you to belie oneself for truth's sake? "Be cunning as serpents and harmless as doves"; so did our Lord Jesus Himself speak to His disciples. You are no serpent, you never belie or constrain yourself. Thus you really serve, not truth, but only yourself; that is to say, your own purity and inviolateness. But I see not only myself, I see also my

88

relations with those among whom I must live in the determination to serve truth with the aid of adroitness and every shrewd device. . . . And I think it is a better imitation of Christ to work among men, even though subject to the protection of princely favor, than merely to dream away one's life in ecstasy and thus to forget all labor and vexations.' "

6. *Einstein as a Professor*

Has Einstein always been a good teacher? Did he like the profession? Very different opinions on these points can be obtained by asking people who have been his students or colleagues.

He had two chief characteristics that made him a good teacher. The first was his desire to be useful and friendly to as many as possible of his fellow beings, especially those in his environment. The second was his artistic sense, which impelled him not only to think out a scientific train of thought clearly and logically, but also to formulate it in a way that gave him, and everyone who listened to him, an æsthetic pleasure. This meant that he liked to communicate his ideas to others.

On the other hand, tending to inhibit these qualities, was the trait that has always been so characteristic of Einstein. I have already mentioned his aversion to entering into very intimate personal relations with other people, a trait that has always left Einstein a lonely person among his students, his colleagues, his friends, and his family. To this was added an absence of ordinary academic vanity. For many professors the reflection of their own personality in so many young people, all of whom repeat what the teacher says, offers a kind of multiplication of their personality. This human characteristic, which may appear as a weakness to some people, is also an asset in the teaching profession. It often leads to a devotion on the part of the pedagogue to his job of teaching that appears selfless and even self-sacrificing. Even though in the last analysis it is a desire for self-expression, the teacher must surrender much of his personality. He must spend a good deal of his own life in serving his students. Einstein did not have this vanity, nor did his personality require multiplication, and consequently he was not ready to sacrifice so much for it. For this reason, too, his relation to his students was likewise ambivalent, but in a very peculiar way.

This way is very obvious from his manner of lecturing. When Einstein had thought through a problem, he always found it necessary to formulate this subject in as many different ways as possible and to present it so that it would be comprehensible to people accustomed to different modes of thought and with different educational preparations. He liked to formulate his ideas for mathematicians, for experimental physicists, for philosophers, and even for people without much scientific training if they were at all inclined to think independently. He even liked to speak about subjects in physics that did not directly concern his discoveries, if he had thought up a method of making these topics comprehensible.

In view of this trait, one might think that Einstein was bound to be a very good lecturer and teacher. Indeed, he frequently was. When he was interested in a subject for scientific, historical, or methodological reasons, he could lecture so that his listeners were enthralled. The charm of his lectures was due to his unusual naturalness, the avoidance of every rhetorical effect and of all exaggeration, formality, and affectation. He tried to reduce every subject to its simplest logical form and then to present this simplest form artistically and psychologically so that it would lose every semblance of pedantry, and to render it plastic by means of appropriate, striking pictures. To these qualities were added a certain sense of humor, a few good-natured jokes that hurt no one, and a certain happy mood mixed with astonishment such as a child feels over its newly received Christmas gifts.

Nevertheless, it was rather irksome for him to give regular lectures. To do so requires that the material for an entire course shall be so well organized and arranged that it can be presented interestingly throughout the year. It means that the lecturer has to interest himself as much in each individual problem as Einstein did in the problems on which all his energy was concentrated. The lecturer must devote himself completely to the material that he is to discuss, and consequently it is very difficult to find time to devote to one's own research. All creative activity requires a great deal of reflection and contemplation, which a superficial observer would regard as a useless waste of time.

There are teachers, especially in German universities, who have arranged their time so precisely that they are able to work out their lectures to the most minute detail and still find time for their own research. But as a result their time is so occupied that they have no place for the unforeseen, for an idea not directly connected with science or the teaching profession, for reflection, or for a conversation with an unexpected visitor. They become dry; any creative and imaginative qualities that they may have are utilized in their scientific research or in teaching students. In daily intercourse they often remind one of squeezed-out lemons and are unable to say anything interesting in company. Such scientists are not infrequent and are found even among the outstanding ones, although they are rare among the truly creative men.

Einstein was always the very opposite of this type. He did not like to grind out information for the students, but preferred to give abundantly of what interested and concerned him. For this reason he put the emphasis on his present field of interest. Also he had too much of an artistic temperament to solve the difficulty of giving a course of lectures in a wide field by the simple method of basing them on a single good book. It was also impossible for him to accumulate enough intellectual energy for his lectures to imbue them all with his spirit. As a result his lectures have been somehow uneven. He has not been a brilliant lecture-room professor, capable of maintaining the same level of interest and excellence in his lectures for an entire year. His single lectures before scientific societies, congresses, and wider audiences, however, were always imbued with a high degree of vitality and left a permanent impression on each listener.

---

7. Generalization of the Special Theory of Relativity

In Zurich and Prague Einstein worked on the solution of questions which were raised by his theory of relativity (Bern 1905). According to the Newtonian principle of relativity, the velocity of a laboratory cannot be determined from observations on the motion of objects within it. Einstein had in 1905 generalized this principle to include optical phenomena, so that observations of neither material bodies nor light rays enable one to determine the velocity of one's laboratory. All this is true, however, only if the motion occurs along a straight line with constant speed. But it is quite consistent with Einstein's theory as developed so far to say that one can determine from experiments in a laboratory L whether it moves with varying velocity relative to an inertial system F. It would thus be possible to learn something about the motion of the laboratory as a whole from the experiment carried out in L. While the velocity itself could not be determined, the changes in speed and direction (acceleration) could be found. Einstein regarded this situation as very unsatisfactory. Ernst Mach had made a suggestion for the correction of this situation by assuming that from the observation in L one does not determine the acceleration relative to an imaginary inertial system, but relative to the fixed stars. Then the events in L would be influenced by actual physical bodies, the fixed stars. Mach's suggestion, however, remained only a program. It was never developed into a physical theory that would enable one to calculate in detail what observable consequences result from the influence of the fixed stars on the observable events in L. It was Einstein's aim to close this gap.

He took as his point of departure the following question: What does Newtonian physics assert about the possibility of learning from experiments carried out in a moving laboratory L whether this room as a whole experiences a change in velocity relative to an inertial system? We have already seen that when the system L is an inertial system, the two Newtonian laws of motion, the law of inertia and the law of force, are valid relative to it. On the basis of daily experience we can likewise see quite easily that these laws *no longer* hold true for L if it is accelerated relative to an inertial system.

For instance, let L be a moving railroad car. If the law of inertia is valid for L, it means that when I am standing in the car, I can remain standing for any length of time at the same spot relative to the car without exerting any force. Experience teaches us, however, that this is only true as long as the car moves along a straight line at a constant velocity. When the car stops suddenly, I shall fall down unless I make a special effort to remain erect. The same thing happens when the car increases its velocity suddenly or rounds a curve. As long as the change in velocity persists, I must make an effort to remain upright. When the velocity becomes constant again, I am able to stand without any effort. This shows that the force that I must exert to remain standing permits me to recognize whether my car L is or is not an inertial system. Moreover, even this crudest kind of experience shows me that the more sudden the stoppage of the car, the greater the required force. More generally speaking, the greater the acceleration, the greater the required force.

From these crude reflections, we can easily develop a method of determining the acceleration (a) of a laboratory L by observing the motion of objects relative to the walls of L. Let us consider, for instance, a little cart lying on the floor of L and free to move in any direction. As long as the laboratory moves in a straight line with uniform velocity, the cart will remain at rest in L, but if the laboratory suddenly changes its velocity, the cart will move with respect to the walls of L as if it had received a jolt. The acceleration (a_0) of the cart due to this recoil as seen in L will be such that its magnitude equals that of a but will be in the opposite direction. For the cart, as described with respect to the inertial system F (in which L has the acceleration a), is a free body not acted on by any force; and hence by the law of inertia its motion is in a straight line with constant velocity. On the other hand, the acceleration of the cart as described relative to F is also equal to the sum of the acceleration (a_0) of the cart with respect to L, and a of the laboratory L itself with respect to F. Since the resulting acceleration must be zero, we have $a_0 + a = 0$. And from this follows $a_0 = -a$, as stated above. Thus the observation of the acceleration (a_0) of the cart in L produced by the motion of L enables us to calculate the acceleration (a) of the laboratory L with respect to the inertial system F.

In the above consideration the cart was initially at rest in the laboratory, but this is not necessary, and in fact it would be even simpler to have it move initially in a straight line with constant velocity in L. Then when the recoil occurs, the cart will in general be deflected from its straight path and move in a curve. From the observation on the shape of this curve we can determine the acceleration of the laboratory.

Furthermore, the acceleration of the laboratory need not be restricted to increase or decrease in its speed. The laboratory may rotate about a certain axis. Such a case is familiar to everyone in the form of a merry-go-round or a railroad car rounding a curve. Just as a recoil in the opposite direction to the acceleration of L occurs in the former case, so in the latter case an impulse directed away from the axis of rotation appears in L. This acceleration is known to physicists as "centrifugal acceleration," and it is entirely analogous to the recoil that occurs when a vehicle begins to move or stop.

In elementary mechanics this situation should be stated as follows: The motion of a body relative to an accelerated or a rotated laboratory cannot be calculated merely from the effect of the gravitation or electric forces acting on it. Accelerations due to recoil and centrifugal forces also occur and must be taken into account. It is often said that these accelerations are due to the appearance of "inertial forces" under such circumstances. They are so called because they arise from the inertia of masses relative to an inertial system.

With Einstein's generalization of the Newtonian principle of relativity to include optical phenomena, it should be possible to use light rays instead of a material object (such as a cart) to find out the acceleration of a laboratory. If a beam of light is arranged so that the rays are parallel to the floor of the laboratory while it is not accelerated, then when it is accelerated the rays will no longer be in a straight line parallel to the floor, but will be deflected. Observations on the magnitude of this deflection will enable us to calculate the acceleration of the laboratory.

Thus we see that according to nineteenth-century mechanics and Einstein's theory of light and motion, advanced in 1905, the acceleration of a laboratory L with respect to an inertial system F has measurable influence on physical occurrences in L, even though it is not possible to state under what observable conditions a system F is an inertial system, But then the part played by the inertial system is none other than that of Newton's "absolute space."

---

8. *Influence of Gravity on the Propagation of Light*

It was Einstein's aim to eliminate this "absolute space" from physics. This did not seem to be an easy task, in view of the fact that such clearly perceptible phenomena as recoil and centrifugal force in railroad cars could not be explained except by the effect of absolute space. Einstein's theory of relativity of 1905 was restricted to motions in a straight line with constant speed and had done nothing in this direction. A new idea leading to even more profound changes had to be introduced into physics. As so often happens, the difficulty was solved by recognizing that it is related to another previously unsolved problem. When one observes the motion of a cart or the deflection of a light ray in a laboratory, the accelerations actually seen may be due to another cause than to the acceleration of the laboratory itself. They may be due to real forces that act on the cart or light ray and, in accordance with Newton's law of force, impart acceleration. How are we to distinguish the effects that arise from this entirely different cause? For forces delivered directly by human beings or some mechanical device, the distinction can be made in this way: Consider two carts of unequal masses instead of one.

If the same force acts on the two, since Newton's law of force states that the change in momentum — that is, the change in the product of the mass and the velocity — is equal to the applied force; the lighter cart will experience a bigger acceleration than the heavier one. On the other hand, if the accelerations are due to inertial forces, they will both be the same. Thus there is this difference: Accelerations due to actual forces (like push or pull) depend on the mass of the object moved; while accelerations due to recoil and centrifugal forces are independent of the mass.

Einstein noticed, however, that there is one type of "real" force that imparts the same acceleration to all bodies. This is the force of gravity. Since the time of Galileo we have known that, apart from the effects of air friction, all bodies fall at the same rate no matter what their masses are. Newton did not regard this as in any way inconsistent with his own law of motion. He simply assumed in his law of universal gravitation that the force of gravity acting on a body is proportional to its mass. The force of gravity acting on any body on the surface of the earth is its weight in the usual terminology. If we denote this by the symbol W, then Newton's assumption can be expressed mathematically as $W = Mg$, where M is the mass of the object and g is a constant at a certain point on the earth. Now, Newton's law of force states that this force Mg is equal to the rate of change of the momentum, which is simply the mass times the acceleration Ma. Thus $Mg = Ma$, and consequently the mass cancels out and we have simply $a = g$. The acceleration due to gravity is independent of the mass and has the same value (g) for all bodies which is, again, Galileo's result.

Einstein realized that this special character of the force of gravity made it impossible to determine the acceleration with which a laboratory moves relative to an inertial system. When we observe in a laboratory a cart executing an accelerated motion, we have no way of deciding whether this is due to the acceleration of the laboratory system as a whole or to a gravitational attraction caused by bodies whose presence may be unknown to us. Into this gap Einstein penetrated with his keen logical analysis, and laid the foundations for a reconstruction of mechanics. As in his earlier paper of 1905, he again related the motion of bodies to the propagation of light, and in 1911 he published a paper entitled *"Über den Einfluss der Schwerkraft auf die Ausbreitung des Lichtes"* ("The Influence of Gravity on the Propagation of Light").

Einstein started out from the following consideration: In a

laboratory *L* that, like an elevator, can move vertically up or down, experiments are performed to observe the motion of objects relative to it. If the laboratory is held by some means such as a cable so that it is at rest with respect to the earth, any object *B* falls downward with the acceleration of gravity, no matter what its mass is or what it is made of. If, however, the laboratory itself is allowed to fall freely owing to the action of gravity, then no object *B* will have any acceleration relative to the laboratory. Everything would occur as if there were no force of gravity. By observing motions with respect to *L* one is not able to decide whether *L* is an inertial system with a field of gravity or whether there is no force of gravity but the laboratory is falling freely. To express the result more generally: it is not possible to distinguish by means of mechanical experiments carried out in a laboratory the accelerations that arise from inertial forces and those that arise from gravitational forces.

To Einstein this conclusion was analogous to Newton's statement that in no case can the speed of rectilinear uniform motion of a laboratory with respect to an inertial system be determined from mechanical experiments within the laboratory. In 1905 Einstein had extended this principle to include optical experiments. In a similar manner he now extended the properties of accelerated motions of objects to include optical phenomena. Thus Einstein advanced the hypothesis that it is impossible, even by means of observations on the rays of light, to determine whether a laboratory is an accelerated system or whether it is at rest or in uniform motion and subjected to a gravitational field. Einstein called this "the principle of equivalence of gravitational forces and inertial forces," or, in short, the *equivalence principle*.

With this principle Einstein was able to predict new optical phenomena that could be observed and hence give an experimental check on the validity of this theory. According to ordinary Newtonian physics, gravity has no effect on the path of a light ray, but according to the equivalence principle, gravitational forces can be replaced by an accelerated motion. The latter, however, as mentioned in the previous section, certainly has an effect on a beam of light. A ray parallel to the floor of a non-accelerated laboratory is no longer parallel when the system is accelerated. Hence Einstein concluded that the path of a light ray is deflected in a gravitational field. The amount of deflection turned out to be very minute because of the enormous velocity of light and no terrestrial experiment is feasible, but Einstein suggested that the effect might be observable for the light that comes to us from the fixed stars and passes near the surface of the sun. In this case the force of gravity is not uniform with the same strength and direction everywhere, but emanates from the center of the sun with a force that decreases in strength as the distance from the surface increases. But Einstein concluded that there would be a deflection in a direction that bends the light ray toward the sun. Since no stars are visible near the sun under ordinary conditions, however, owing to the blinding sunlight, Einstein pointed out in his paper that:

"Since the fixed stars in the parts of the sky near the sun become visible during a total eclipse, it is possible to check this theoretical conclusion by experiment."

By assuming that the force of gravity has the value accepted by Newton, Einstein showed by a very simple calculation based on his *equivalence principle* that a ray of light coming from a fixed star and just grazing the border of the sun will be deflected from its straight path by 0.83 seconds of an arc. Consequently, if one photographs the fixed stars near the sun during a total solar eclipse and compares their positions with those where the sun is not near them, differences between their positions are to be expected. Since the light rays are bent toward the sun, the stars must appear shifted away from it, the magnitude depending on the proximity of the rays to the sun as they pass by it. Einstein concluded his paper with these words:

"It would be extremely desirable if astronomers would look into the problem presented here, even though the consideration developed above may appear insufficiently founded or even bizarre."

No matter what one may think of Einstein's hypothesis, he had brought forward a definite observational check on his theory. Since total solar eclipses are not very frequent and are observable only from a very limited part of the earth, astronomers were stimulated to undertake interesting and adventurous journeys. It took three years, however, until 1914, to find enough support and money to dispatch an expedition equipped to perform this observation. But just as this first expedition left Germany for Russia, World War I broke out, and the members of the expedition became Russian prisoners and were prevented from making the observation.

9. Departure from Prague

While he was a professor at Prague, Einstein not only founded his new theory of gravitation but also developed further the quantum theory of light that he had begun while in Bern. His hypothesis that a quantum of violet light possesses much more energy than that of red light seemed to be in agreement with experimental results on the chemical action of light. Every photographer is familiar with the fact that the action of violet light is much stronger than that of red light on a photographic plate. Einstein started with the simple assumption, very closely related to his photon theory of light, that the chemical decomposition of a molecule always takes place with the absorption of only a single light quantum. In his paper published in 1912 under the title *"Über die thermodynamische Begründung des photochemischen Äquivalenzgesetzes"* ("On the Thermodynamic Foundations of the Photochemical Equivalence Law), he showed that the assumption is also in accord with the general principles of thermodynamics.

About this time, however, Einstein began to be much troubled over the paradoxes arising from the dual nature of light: the wave character exemplified by the phenomena of interference and diffraction and the particle aspect shown by the photoelectric and chemical actions. His state of mind over this problem can be described by this incident:

Einstein's office at the university overlooked a park with beautiful gardens and shady trees. He noticed that there were only women walking about in the morning and men in the afternoon, and that some walked alone sunk in deep meditation and others gathered in groups and engaged in vehement discussions. On inquiring what this strange garden was, he learned that it was a park belonging to the insane asylum of the province of Bohemia. The people walking in the garden were inmates of this institution, harmless patients who did not have to be confined. When I first went to Prague, Einstein showed me this view, explained it to me, and said playfully: "Those are the madmen who do not occupy themselves with the quantum theory."

Soon after Einstein's arrival in Prague, he had received an offer of a professorship of theoretical physics at the Polytechnic School in Zurich, the institution from which he had graduated.

The Polytechnic belongs to the Swiss Confederation and is a larger and more important institution than the University of Zurich, where Einstein had first taught and which is maintained by the canton of Zurich. Einstein was in doubt whether or not to return to Zurich, but his wife decided the matter. She had never felt at ease in Prague and was attached to Zurich, which had become her ideal home while she was a student there.

Einstein informed the university at Prague that he would leave it at the end of the summer semester of 1912. But with his usual indifference to all official formalities, he did not send to the administrative authorities the documents that had to be filled out when one resigned from the service of the Austrian state. The Ministry of Education in Vienna did not receive the application that had to be forwarded in such cases. One can well imagine that the official in charge was unhappy at being unable to close Einstein's record according to regulations. For many years the dossier for the "Einstein case" remained incomplete in a pigeonhole. Some years later, when Einstein went to Vienna for a lecture, a friend told him that the official in the ministry was still unhappy over the gap in the records. Einstein with his good nature did not want to make anybody unhappy. He visited the ministry, made his excuses to the official, and filled out the appropriate form. The pigeonhole lost its blemish.

Einstein's sudden departure from Prague gave rise to many rumors. An editorial in the largest German newspaper of Prague asserted that because of his fame and genius Einstein was persecuted by his colleagues and compelled to leave the city. Others maintained that because of his Jewish origin he had been badly treated by the administrative authorities in Vienna and therefore did not want to remain in Prague any longer. Einstein was much astonished by all this talk, as his stay in Prague had been a very pleasant one, and he had been favorably impressed by the Austrian character. Since he did not like to create any unpleasantness for anyone, he wrote a letter to the head of the Austrian university administration in Vienna. Before taking over my position in Prague, I paid a visit to this man. He was a Pole and embraced me according to the Polish custom as if I were a close friend. In the course of the call he told me about Einstein's letter and said with great enthusiasm: "I received a splendid letter from Mr. Einstein, such as one is not accustomed to receive from a professor of our universities. I recall this letter very often. It gave me a great deal of satisfaction, particularly since so many attacks were directed against our government on account of Einstein."

For me Einstein's departure from Prague is bound up with a rather humorous story, which I wish to relate because it is linked with the checkered history of our time. Like every Austrian professor, Einstein had had to get a uniform. It resembled the uniform of a naval officer and consisted of a three-cornered hat trimmed with feathers, a coat and trousers ornamented with broad gold bands, a very warm overcoat of thick black cloth, and a sword. An Austrian professor was required to put on this uniform only when taking the oath of allegiance before assuming his duties or when he had an audience with the Emperor of Austria. Einstein had worn it only once, on the former occasion. Since the uniform was rather expensive and he had no use for it after his departure, I bought it for half the original price. But before Einstein gave me the uniform, his son, who was then perhaps eight years old, said to him: "Papa, before you give the uniform away, you must put it on and take me for a walk through the streets of Zurich." Einstein promised to do so, saying: "I don't mind; at most, people will think I am a Brazilian admiral."

I too wore it only once, when taking the oath of allegiance, and I had it in my trunk for a long time. After six years the Austrian monarchy disappeared and the Czechoslovakian Republic was established at Prague. The oath of allegiance to the Emperor was replaced by that of allegiance to the Republic, and the professors had no uniform any more. The uniform remained only as a memory of Franz Joseph and Einstein. Soon after the Russian Revolution, when a large number of refugees, many of whom were officers, came to Prague, my wife said to me: "Why should such a good coat lie unused when so many are freezing? I know a former commander in chief of the Cossack army who cannot buy a warm winter coat. Einstein's coat looks almost like the coat of a high-ranking cavalry officer. It will please the general and keep him warm." We gave him the coat, but he was not interested in its distinguished past. The rest of the uniform, including the sword, remained in the German University. When the Nazis invaded Czechoslovakia in 1939, the university became a bulwark of Nazism in the east and Einstein's sword probably became booty of a Nazi soldier, a symbol of the final defeat of "international Jewish science" — until 1945, when the Red Army entered Prague.

V

EINSTEIN AT BERLIN

1. The Solvay Congress

In the fall of 1912 Einstein entered upon his duties as professor at the Polytechnic in Zurich. He was now the pride of the institution where he had once failed to pass the entrance examination, where he had studied and met his wife, and where on graduation he had been unable to obtain even a minor position.

As early as 1910, when Lampa was considering Einstein's appointment to Prague and seeking an opinion of his qualifications from a scientist who was generally recognized as an authority, Max Planck, the leading theoretical physicist, had written to the faculty committee at Prague: "If Einstein's theory should prove to be correct, as I expect it will, he will be considered the Copernicus of the twentieth century." Einstein was already beginning to be surrounded by an aura of legend. His achievements were characterized as a turning-point in physics comparable to the revolution initiated by Copernicus.

In 1911, when a conference of a small number of world-famous physicists was to be convened in Brussels to discuss the crisis in modern physics, there was no question that an invitation would be extended to Einstein. The selection of the conferees was suggested by Walter Nernst, a leading investigator in the fields of physics and chemistry, and among others there were Sir Ernest Rutherford of England, Henri Poincaré and Paul Langevin of France, Max Planck and Walter Nernst of Germany, H. A. Lorentz of Holland, and Madame Curie of Poland, who was working in Paris. Einstein represented Austria, together with Friederich Hasenöhrl, the Viennese, whose name after his tragic death was to be linked in a peculiar manner with the fight against Einstein. This conference was Einstein's first opportunity to meet these scientists whose ideas shaped the physical research of this period.

The costs of the conference, including the traveling expenses to Brussels and the living expenses there and in addition a remuneration of a thousand francs to each conferee, was defrayed by a rich Belgian named E. Solvay. This man had been successful in the chemical industry, but his hobby was a physical theory of the outmoded mechanistic type. Although it led to many complications and not to the discovery of new laws, he was greatly interested in attracting the attention of physicists to his theory and in learning their opinions about it. The clever chemist Walter Nernst, who was in social contact with him thought that this rich man's hobby might be utilized for the benefit of science while at the same time fulfilling Solvay's desire. He proposed that he call a conference of leading physicists to discuss the present difficulties in their science, to whom he could present his ideas on this occasion. The conference became known as the Solvay Congress. In the opening address Solvay presented a summary of his ideas, and the conferees then discussed the new developments in physics. Finally in the concluding address Solvay thanked the speakers for their interesting discussions, emphasizing how much pleasure he had derived from them. Nevertheless, all this had not shaken his faith in his own theory. All the speakers had avoided entering upon any criticism of his theory, to prevent any conscientious scruples arising between their feelings of gratitude and courtesy toward their host and their scientific convictions. Solvay was imbued with such sincere interest in the advancement of science that he subsequently convened similar conferences quite often, and at these meetings Einstein always played a leading role. A man like Nernst who has the interests of science at heart and is practical can utilize such opportunities for the benefit of progress in scientific research.

The world marveled at the great number of new and astonishing ideas and at the thoroughness with which these concepts were developed, presented, and arranged in a larger chain of ideas that Einstein had already produced in 1912 after less than ten years as a physicist. But Einstein himself thought only of the defects and the gaps in his creations. His new theory of gravitation, which he had made public in 1911 at Prague, dealt only with one very special case of the effects of gravity. Only the case where the force of gravity has the same direction and intensity throughout the entire space under consideration was completely clear, and the theory as developed so far was unable to furnish a complete solution to cases where the force of gravity had different directions at different points in space.

Up to this time Einstein had solved his problems with the

simplest mathematical aids and had looked upon every exaggeration in the use of "higher mathematics" with the suspicion that it was not due to any desire for clarity, but rather to dumbfound the reader. Now a new trend appeared in his work. It has been mentioned that while in Prague Einstein had already felt that the development of a still more general theory required more complicated mathematical methods than those he had at his command. He had discussed this matter with his colleague Pick, who had called his attention to the new mathematical theories of the Italians Ricci and Levi-Civita. In Zurich Einstein found among his colleagues his old friend Marcel Grossmann, and with him he now studied these new mathematical methods. In collaboration with him Einstein succeeded in preparing a preliminary sketch of a general theory of gravitation in which every case of the action of the force of gravity was contained. This work, published in 1913, still contained many defects, however, and they were not removed until the complete theory was finally published during the World War. We shall discuss it in detail later.

---

2. *Trip to Vienna*

In the fall of 1913, at the Congress of German Scientists and Physicians in Vienna, Einstein was invited to present a summary of his new ideas on the theory of gravitation. Even then he was regarded as an unusual phenomenon among the physicists, and it was rumored that he had "thought up" a general theory of relativity which was "even more incomprehensible" than his special theory of 1905 and even further removed from the physics of the laboratory. In consequence a large audience crowded the room where he was to speak. Einstein, however, took the most obvious and easily understood ideas as his points of departure and tried step by step to awaken in his listeners a feeling that radical changes were necessary if only one tried to see clearly the defects and gaps in the previous theories.

His explanation was approximately as follows: At first, investigations of the nature of electricity were concerned only with electrical charges. The forces involved in the mutual attraction and repulsion of these charges were known, and it was also known that, like the Newtonian gravitational forces, they decreased with the square of the distance between the charges. Later, electric currents were discovered. and it was found that they could be generated by moving magnets, as well as by moving electric charges. This led to the industrial application of electricity. Finally, electromagnetic waves were discovered and utilized in wireless telegraphy and radio. No one had imagined that all this would develop from the simple attraction of electrical charges. In the theory of gravitation we are still in this first period where we are acquainted only with the law of attraction between material bodies. We must create a theory of gravitation that will be as far removed from the simple Newtonian theory of attractions as the theory of radio waves is from the views of Benjamin Franklin.

In his lecture Einstein also mentioned that previous to his work a young Viennese physicist had already developed some of the mathematical ideas that he had used in his theory. He asked whether this man was in the audience, as he did not know him personally. And in fact a young man rose and Einstein asked him to remain standing so that the entire audience could see him. This man was Friedrich Kottler, later employed by the Eastman Kodak Company at Rochester, New York.

Einstein took this opportunity of his stay in Vienna to become personally acquainted with the physicist and philosopher Ernst Mach, who had had such a profound influence on the development of Einstein's ideas. (Ch. II) At the University of Vienna Mach had lectured on the history and theory of the "inductive" sciences — that is, sciences such as physics and chemistry that advance from individual observations to general laws. For more than twelve years, however, Mach had suffered from a severe paralysis and had retired from his position. He lived in his apartment in a suburb of Vienna, and occupied himself only with his studies and receiving occasional visitors. On entering his room one saw a man with a gray, unkempt beard and a partly good-natured, partly cunning expression on his face, who looked like a Slavic peasant and said: "Please speak loudly to me. In addition to my other unpleasant characteristics I am also almost stone-deaf." Mach was very much interested in meeting the originator of the new theory of relativity.

Even though Einstein greatly admired Mach's ideas on the logical structure of physics, there were many things he could not accept. According to Einstein's judgment Mach did not give enough credit to the creative mind of the scientist who imagines general laws beyond a mere economic description of facts.

Mach's opinion, that the general laws of science are only a means by which individual facts can be remembered more easily, did not appear satisfactory to Einstein. To him the phrase "remembered more easily" could in this connection apparently mean only "remembered with less effort." Mach's *economy* seemed to be *economy* in a psychological sense.

Hence, after conversing awhile with Mach, Einstein raised the following question: "Let us suppose that by assuming the existence of atoms in a gas we were able to predict an observable property of this gas that could not be predicted on the basis of a non-atomistic theory. Would you then accept such a hypothesis even if the calculations of its consequence required very complicated computations, comprehensible only with great difficulty? I mean, of course, that from this hypothesis one could infer the interrelation of several observable properties that without it would remain unrelated. It is then 'economical' to assume the existence of atoms?"

Mach answered: "If with the help of the atomic hypothesis one could actually establish a connection between several observable properties which without it would remain isolated, then I should say that this hypothesis was an 'economical' one; because with its aid relations between various observations could be derived from a single assumption. Nor should I have any objection even if the requisite computations were complicated and difficult."

Einstein was exceedingly satisfied with this statement and replied: "By 'simple' and 'economical' you mean, then, not a 'psychological economy' but rather a 'logical economy.' The observable properties should be derived from as few assumptions as possible, even though these assumptions appear 'arbitrary' and the computation of the results might be difficult."

With *economy* interpreted in this *logical* sense, there was no longer any conflict between Mach's standpoint and Einstein's as to the criteria to be filled by a physical theory. Although Mach made the concession in conversation, yet Einstein saw in his writing only a demand for "psychological economy." Thus for the moment Einstein was satisfied, but he retained a certain aversion to the "Machist philosophy."

3. *Invitation to Berlin*

Einstein's fame had grown so great by now that many centers of scientific research desired to secure him as an associate. For several years efforts had been exerted to develop Berlin not only as a center of political and economic power but also as a center of artistic and scientific activity. Emperor Wilhelm II, who liked to associate with Americans, had learned from them that in the United States there were, in addition to the universities, institutions devoted solely to research, to which rich business men such as Rockefeller, Carnegie, and Guggenheim donated large sums of money. The Kaiser was aware that military and economic power required as a basis an organization of scientific research, and he wanted to use his influence to found similar research institutions in Germany. For his purpose this was particularly important in the fields of physics, chemistry, and their applications.

To further these aims Wilhelm II founded the Kaiser Wilhelm Gesellschaft, within which rich industrialists, merchants, and bankers united to help build research institutes. The members received the pompous title of "senator" and the right to wear a handsome gown, and they were sometimes invited to breakfast with the Emperor — an invitation that cost them each time a great deal of money. During the course of the conversation at these breakfasts, the Emperor would mention that money was requested for a particularly important field of research.

The erection of these institutes had the additional advantage that scientists whom the government did not want to appoint as professors in the universities, because of pedagogical, political, or other reasons, could still be employed in a way beneficial to the German Reich. Men of outstanding eminence were sought for these institutes, and the appointments could be made solely on the basis of scientific achievement.

Owing to the fact that the Kaiser was interested not only in physics and chemistry but also in modern Biblical research, the first president of the Kaiser Wilhelm Gesellschaft was the liberal Protestant theologian Adolf Harnack. He was persuaded by Max Planck and Walter Nernst to invite Albert Einstein, the rising star in physics, to Berlin.

Planck and Nernst, the leaders in German physics at this time, were to play important roles in Einstein's life. They represented two very different types of German scientist. Max Planck was a member of a Prussian family of military officers and government officials. He was tall, slim, an enthusiastic mountain-climber, and a lover of classical music. Basically, he accepted the philosophy of his social class; he believed in the mission of the Kaiser to make the world happy with his conception of German culture, and in the right of his class to provide the leaders for Germany and to exclude people of other origins from such functions. On the other hand, he was an ardent adherent of Kantian philosophy in the diluted form in which it had become the common religion of the German academic and governmental circles. He believed with Kant in the duty of doing everything that is "qualified to become a general rule of human conduct." He also believed in the international mission of science and in non-German co-operation with Germans in scientific research. But since his immediate emotional reaction was to respond in terms of the philosophy of the Prussian bureaucracy, an appeal to his reason was necessary to make him recognize the rights of aliens. As he was conscientious and an idealist, such an appeal was usually successful.

On the other hand, Walter Nernst, although a great scientist and scholar, exhibited the mentality characteristic of a member of the merchant class. He had no national or class prejudice and was imbued with a type of liberalism that is often peculiar to business men. He was short, active, witty, and quick of apprehension. He occasionally utilized his craftiness in professional life, and his students jokingly referred to him as the *"Kommerzienrat,"* a title conferred in Germany on successful business men. There was a story about him that he was the only physicist who had ever signed a contract with an industrial firm in which the advantage was not on the side of the firm. This contract concerned his invention of an electric light bulb, widely known for a time as the Nernst lamp. He earned a good deal of money from it, but the lamp soon fell into disuse.

Planck and Nernst went to Zurich personally to influence Einstein in favor of their plan. This was the following: There was as yet no separate research institute for physics and no hope that any such institute could be built in the near future. Nevertheless, Einstein was to become the director of the institute that was being planned, and in the meantime to assist in a consultative capacity in the physical research being carried on in other institutes. In addition he was to become a member of the Royal Prussian Academy of Science. To be a member of this body was considered a great honor; and many outstanding professors at the University of Berlin never succeeded in achieving it. Although membership in the Academy was only an honorary position for most of the incumbents, a few were endowed by foundations which paid a sufficiently large salary. Such a position was offered to Einstein. Both in the Academy and in the Kaiser Wilhelm Institute his main occupation was to be the organization of research. He was to have the title of professor at the University of Berlin, but unencumbered by any obligations or rights, except that of lecturing as much or as little as he desired. He was to have nothing to do with the administration of the university or with examinations, or in the appointment of new professors.

There were great advantages offered by this invitation. Besides the academic honors that the Prussian Academy bestowed on him, it meant that he would receive a much larger salary than he did at Zurich. He would also be in a position to devote himself entirely to research and would have as much of an opportunity as he desired to come into contact with the many leading physicists, chemists, and mathematicians who were in Berlin. Despite his unusual talents he could still expect to be stimulated by new ideas, since it is always fruitful to receive the criticism of so many scientists capable of independent thinking, working in many different fields. In addition it meant that Einstein would not be obliged to give regular lectures, which he considered very burdensome.

On the other hand, it was difficult for him to decide to return to the center of that Germany from which he had fled as a student. It seemed to him even a kind of betrayal of his convictions to become a member of a group with which he did not harmonize in so many respects, simply because it was connected with a pleasant position for himself. It was for him a struggle between his personality as a scientific investigator who could benefit by moving to Berlin, and his feeling as a member of a certain social group.

In addition there were also personal factors that entered into the decision. Einstein had an uncle in Berlin a fairly successful businessman, whose daughter, Elsa, was now a widow. Einstein remembered that his cousin Elsa as a young girl had often been in Munich and had impressed him as a friendly, happy person. The prospects of being able to enjoy the pleasant company of this cousin in Berlin enabled him to think of the Prussian capital somewhat more favorably. And so Einstein finally decided to accept the offer, and at the end of 1913 he left Zurich.

4. *Einstein's Position in the Academic Life of Berlin*

Soon after his arrival in Berlin Einstein separated from his wife, Mileva, with whom in many respects he was no longer in accord, and he now led a bachelor life. When he became a member of the Academy he was just thirty-four years old. He was a young man among men who were in general much older, men with proud pasts and great authority, and many of them of great achievements also. The feeling of strangeness that he felt there from the beginning, however, arose only in a very small part from the difference in age. Most of these men were, so to speak, "veterans of university life." Everything that happened in these circles seemed to them extraordinarily important, and election to the Academy appeared to be the culmination of their aspirations. All this could not make any great impression on Einstein, who was already on his way to world fame before he was a great man in narrower academic circles.

While Einstein was still in Zurich and a long time before the invitation from Berlin, someone happened to remark in his presence: "It is really a pity that no one ever enters the Academy while still a young man, at a time when it would still make him happy." "If that is the case," Einstein rejoined, "I could be elected to the Academy immediately, since it would not make me happy even now."

In an academy there is always much that is comical. Einstein appreciated this as much as he had the comedy in the faculty meetings at the University of Prague. Moreover, the comical aspects of such bodies are hardly to be avoided. This is due to the circumstance that even the greatest scientists of the country must deal with questions that are often of but slight significance, but that must be discussed with the same thoroughness and earnestness as if they were scientific questions of prime importance. For instance, whether a work to be published by the Academy is to be done in two or three volumes, whether A should receive one hundred marks for his work and B one hundred and twenty or vice versa, and many similar questions were discussed with considerable acumen and temperament. Also, in accordance with an old tradition, the papers that were to be printed in the transactions of the Academy had to be presented, even if only in summary form when the Academy met. Since as a rule these papers dealt with very special subjects, they were completely incomprehensible and uninteresting to most of the members. One described a rare moss found in a certain part of Finland, another discussed the solution of a complicated mathematical equation, and still another the deciphering of a Babylonian inscription that could be read only with difficulty. In order to be polite, one had to show a certain interest; actually the members often had to make an effort not to fall asleep during the sessions. This was all very natural, but the contrast between earnestness and triviality could not but appear comical. Einstein was well able to appreciate this, and his sense of humor made it easier for him to endure much that was unpleasant.

Professor Ladenburg, a German physicist who lived and worked for a long time with Einstein in Berlin and who is now at Princeton University, once said to me: "There were two kinds of physicists in Berlin: on the one hand was Einstein, and on the other all the rest." This is a very good characterization of Einstein's position. To all outward appearances he was a member of a professional group, but he never belonged to its rank and file. His aloofness was always noticeable, and his status may be described aptly by the slang expression: "He was in a class by himself."

Einstein's contradictory attitude toward co-operation with others and his aloofness we have so often noted manifested itself also very definitely in his attitude to his occupation as professor. He frequently expressed the opinion that a scientist should earn his living from a "cobbler's job." If he is paid to discover new theories he must constantly be thinking: "Discoveries cannot be made on order and if I don't discover anything, I shall disappoint my employers and receive my pay for nothing." But if he is active as a technician or a teacher, he is always doing something useful and hence has a clear conscience. On his own ideas he should work for pleasure only.

This may be a little exaggerated, since ultimately pure science has a social value too. But it is certain that Einstein has had a definite aversion to pure research as a profession. Actually, as fate would have it, after moving to Berlin he was permanently what he did not want to be — a pure research worker. This

was the situation in Berlin, and he was to obtain a similar position later in Princeton.

The contradiction in Einstein's relation to his environment is
also manifested in his aversion to giving regular lectures cov-
ering the entire field of physics, even though probably few
physicists are interested in and familiar with more fields of physics than he is. Very many, indeed most specialists, in physics as
well as in other sciences, are hardly able to understand anything complicated that does not belong within their narrow
field. Most of them are inclined to exaggerate enormously the
significance of their "subject"; they consider every thought devoted to outside matters as a betrayal of pure research and a
concession to dilettantism. Einstein has been the exact opposite
of this type. One may recount to him the most complicated
physical theory; he will listen attentively, and through his questions show immediately that he has grasped the essence of the
matter. He will almost always make a good critical comment
or a helpful remark. Even when the construction of some apparatus is being discussed, he concerns himself with every significant detail and intervenes with his advice.

Obviously Einstein is not a "teacher" if this word is taken in
the sense in which it is current in professional circles. On the
other hand, quite in accord with his divided attitude, which we
have previously noted, he is more interested than most professors in social matters, such as the position of scientific teaching
and research in the social life of man. He has always tried to
clarify for himself and others the reciprocal relation between
science on the one hand and society, religion, and international
co-operation on the other.

In Berlin, as in many universities, it was customary to have
a physics colloquium every week where recently published researches were discussed. It gave physicists who worked in different institutions an opportunity to exchange opinions and
ideas about every new discovery and theory. During the period
of Einstein's stay, between 1913 and 1933, the Berlin seminar
was an especially interesting gathering, such as hardly existed
anywhere else in the world. Besides Einstein, Planck, and
Nernst, there were Max von Laue, the discoverer of the diffraction of X-rays by crystals; James Franck and Gustav Hertz,
who discovered that light of specific color can be produced by
the impact of high-velocity electrons; and Lise Meitner, a Viennese girl who had made such great discoveries in the field of
radioactive phenomena that Einstein liked to call her "our

Madame Curie," and in private sometimes expressed the opinion that she was a more talented physicist than Madame Curie herself. During the latter years of this period there was also Erwin Schrödinger, another Austrian, who derived the quantum theory of the atom from a wave theory of matter.

The discussions with such outstanding investigators were of value even to a man with great creative power like Einstein. At the very least he was spared the necessity of reading much that would otherwise have taken up a great deal of time. Einstein attended the colloquium quite regularly and took an active part in the discussions. He liked to branch out into all sorts of problems, and his remarks refreshed all who were present. His questions alone were sufficient to exert a stimulating influence. On such occasions there are always many who are ashamed to ask questions because they do not wish to appear ignorant, and usually it is just the people who take the longest to comprehend something who are shy. Since Einstein could never be suspected of slow comprehending, he did not hesitate to ask questions that would otherwise be considered as naïve. Such "naïve" questions, however, are often very stimulating, because they frequently deal with fundamental problems that no one really dares to touch. Most specialists would like to make believe that they understand the fundamentals and are only seeking to explain secondary matters. Einstein's questions, which very often threw doubt upon a principle that appeared self-evident, gave the seminar a special attraction. After Einstein's departure from Berlin in 1933, the colloquium presented the appearance of a gathering from which the guest who had endowed it with lustrous brilliancy had departed.

5. *Relationship with Colleagues*

Einstein's attitude to the teaching profession was also connected with his concrete relations to his colleagues. There is no doubt that the immediate impression he made on his colleagues was that of a very likable person. He was filled with a simple natural friendliness toward everyone, no matter what the individual's position. He was very amicable toward people of high rank; he had such a feeling of inner security that he did not have to demonstrate his independence by being short with people. He never took part in any intrigues such as occur in

Einstein with Donald Menzel (astronomer) and George Birkhoff (mathematician) on the grounds of the Harvard Observatory. The child is Carl Shapley

Einstein and Rabindranath Tagore

all corporate bodies and professions, including university faculties. He was not considered dangerous to anyone, because he never tried to frustrate anyone's desire. He was ready to converse in a friendly manner about anything and everything; he liked to crack jokes and to laugh at other people's jokes. He always avoided putting himself in the foreground and forcing his will on other people. This would have been easily possible if he had used his personality and his fame, but he did this very rarely, and at most to defend himself against unreasonable demands, never in an offensive manner.

He always managed to maintain a certain "free space" around himself which protected him from all disturbances, a space large enough to contain a world erected by an artistic and scientific imagination.

There were also certain features in his Berlin environment, no matter whether one calls them national or cultural peculiarities, that produced in Einstein a feeling of strangeness and loneliness. In the eighteenth century, under Frederick the Great, Frenchmen such as Voltaire and d'Alembert had been the pride of the Berlin Academy. But since the Bismarck era and the turning of the German intellectuals to nationalism, an atmosphere of voluntary or involuntary submission to the philosophy of the new German Reich had become more and more prevalent, at first under the influence of Bismarck and later under that of Wilhelm II. This sentiment was also connected with a certain emphasis on the superiority of the German nation or race, which, although at that time not yet very conspicuous, was already quite evident to Einstein.

But what particularly annoyed Einstein from the beginning was the cold, somewhat mechanical manner of the Prussians and their imitators, whom Einstein had feared as a student and from whom he had fled. Einstein sometimes expressed his feeling as follows: "These cool blond people make me feel uneasy; they have no psychological comprehension of others. Everything must be explained to them very explicitly."

As a result, for a man who came from a somewhat different environment, and especially for a man like Einstein with a strong intuitive feeling for the significance of human relationships, life among them was often bound up with conflicts. Einstein experienced this feeling of strangeness even in his relations with a man like Max Planck, who had done so much for Einstein's recognition as a scientist, who had supported and worked for his election to the Academy, and who had a high opinion of him as a person. Einstein could never get rid of the feeling that the emotion and ideas of a man like Planck were actually opposed to his own and that it was only by means of rational arguments that Planck forced himself to say or do something in agreement with Einstein's views or intentions. Einstein always sensed the existence of a barrier behind which something hostile lay hidden, and therefore behind which he preferred not to look; but the conviction that this barrier existed produced a sense of uneasiness, which, while sometimes hardly noticeable, was never completely absent.

The degree to which this Prussian reserve and mechanical thinking weighed upon Einstein became evident when Erwin Schrödinger, the Austrian, came to Berlin as Planck's successor. There were no barriers; there was an immediate understanding between the two men without any long explanations, and an agreement on the manner in which they would act toward each other without first having to call upon Kant's categorical imperative.

Einstein's solitary position in academic circles was due also to the fact that he did not like to take part in the problems of professional daily life; he was unable to take them seriously. The daily life of a scholar is often a matter of discussing and becoming excited about the frequency with which his papers are published, which colleagues have or have not published anything, which colleague has frequently or infrequently cited which other colleague, or who intentionally or unintentionally has failed to cite somebody else. There are discussions of the merits of individual professors, the honors that they have or have not received from their own or other universities, and the academies to which they have been elected. Then again the conversation may turn to the number of students for whom the professors have been able to obtain positions, the students and teachers whom they have been able to prevent from obtaining positions, whether they have any influence with superior officials, and whether they are able to obtain money for their department from these authorities.

Taken as a whole, all these problems add up to a tremendous total of interests and intellectual effort, in which Einstein hardly participated. It would be very unjust to maintain that all these conversations are valueless for scientific activity. On the contrary, they have their justification in social life. Nevertheless, too much attention to these details may prevent one from dealing with the actual problems of science. It is perhaps compatible with research in a very specialized field, but it is undoubtedly a considerable hindrance if one looks upon science as a religion and a philosophy that influence one's entire life, as Einstein had done throughout his life. Yet one ought not to overlook the fact that, in consequence of this withdrawal from the more trivial details of the daily life of most professors, he often deprived himself of opportunities of obtaining concrete influence. Every social group is so constituted that petty things are inextricably interlinked with important matters, and as a result, by manifesting an aversion to the petty things, one easily loses the possibility of exercising an influence on more important matters. For a man like Einstein, however, such means of exerting his influence were so disagreeable that he could very rarely decide to make use of them.

This aversion to petty talk was more than compensated for in unbounded readiness to discuss scientific problems and questions of general interest with his colleagues. Without the slightest trace of pomposity he turned to his colleagues for advice, even to those younger than he if they were more conversant with special problems. And all this took place quite informally.

Einstein was always very intent on being a person who did not require any special consideration. On one occasion he was supposed to pay his respects to a member of the Berlin Academy. He was not very fond of such formal visits, but he had heard that Professor Stumpf, a well-known psychologist, was greatly interested in the problems of space perception. Einstein thought that he would be able to discuss matters of mutual interest that might have some connection with the theory of relativity, and he decided to make this call. On the chance that he might find the professor at home, he went there at eleven in the morning. When he arrived, the maid told him that the *Herr Geheimrat* was not at home. She asked Einstein whether he would like to leave a message, but he said it was unnecessary. He did not want to disturb anyone, and would come back later in the day. "In the meantime," he said, "I'm going to take a walk in the park." At two in the afternoon Einstein returned. "Oh," said the maid, "since you were here the *Herr Geheimrat* came home, had his lunch, and because I did not say that you would come back, he is taking his afternoon nap." "Never mind," said Einstein, "I'll come back later." He went for another walk and came back again at four. This time he was finally able to see the *Geheimrat*. "You see," Einstein said to the maid, "in the end patience and perseverance are always rewarded."

The Geheimrat and his wife were happy to see the famous Einstein and assumed that he was now making his formal introductory visit. Einstein, however, immediately began to talk about his new generalization of the relativity theory and explained in detail its relation to the problem of space. Professor Stumpf, who was a psychologist without any extensive mathematical knowledge, understood very little of the discussion and was hardly able to put a word in edgewise. After Einstein had talked for about forty minutes, he remembered that he was actually supposed to be paying an introductory call and that it had already lasted too long. Remarking that it was quite late already, he departed. The professor and his wife were dumbfounded, for they had had no opportunity to ask the customary questions: "How do you like Berlin?" "How are your wife and children?" and so on.

---

6. *Relationship with Students*

Einstein's chief activity in Berlin was conversing with colleagues and students about their own work and advising them about their research programs. He did not have to give regular courses and he lectured only occasionally, either on his own special field or on subjects of general interest to lay audiences.

Even among professors whose chief activity was giving regular courses of lectures, the guidance of students in research was considered to be an important part of their duties. It was the pride of the teachers at the German universities to have as many scientific investigations as possible carried out and published by students under their direction. Hence many students who otherwise would never in their life have produced an independent piece of work published at least a dissertation when they acquired the doctorate. For this purpose a professor had to provide subjects of research for even untalented students who lacked ideas of their own, and then push them along until the studies were completed. In many cases the teacher could have carried out these investigations better and more rapidly if he had done the work himself, and so a certain lack of selfishness was required to waste so much effort on incompetent students.

On the other hand, many professors were themselves not very

talented. They divided the subject on which they were working into innumerable small parts and then let each part be handled by a student. The student's task under these circumstances was relatively easy, and was dealt with in great detail so as to create an impression of being important. In this way there arose what was known in Germany as the *"Betrieb"* (mill), where to all outward appearances no distinction is made between worthwhile ideas and trivialities. Everything produced was a "contribution to the literature," which had to be cited by every subsequent writer if he wanted to be "scientific." An agreeable feeling of activity surrounded both teacher and students. They became so engrossed in this activity and industry that the larger problem that the partial studies were supposed to elucidate was often forgotten. The production of dissertations and papers became an end in itself.

Einstein never evinced any interest in this kind of activity. Above all, he did not like the idea of raising easy questions and preferred to deal only with problems that naturally arise when investigating fundamental bases of natural phenomena. Einstein once remarked about a fairly well-known physicist: "He strikes me as a man who looks for the thinnest spot in a board and then bores as many holes as possible through it." He esteemed most highly those who occupied themselves with difficult problems even if they were able to advance few steps in this thicket, or even if they themselves could not extend our knowledge in any positive sense, but were only able to make clear to the world the magnitude of the difficulties involved. With this conception of scientific work Einstein was not the man to have many students working under him. Whatever he undertook was always so difficult that he alone was able to carry it through.

There was also a great difference between Einstein's attitude and that of his colleagues toward the peculiar, pseudo-scientific questions that university professors often receive by mail from dilettantes of science. Einstein was remarkably patient in answering them, and in many respects it was easier for him than for most other scientists. Many professors, even outstanding ones, are so immersed in their own ideas that it is difficult for them to comprehend ideas that deviate from the traditional, or are merely expressed in a way differing from that commonly used in scientific books. This difficulty frequently manifests itself in hatred or contempt for amateurs since the professors are often actually incapable of refuting the ingenious objections made by dilettantes to scientific theories. As a result they give the impression of incompetence and the falsehood of "academic science." Einstein, on the other hand, did not regard the differences between the layman and the professional as being very great. He liked to deal with every objection and had none of the reluctance that makes such work so difficult for others; and this was especially important in his case since laymen frequently occupied themselves with and discussed the relativity theory.

These characteristic features in his psychological constitution and in the manner in which he carried on scientific research brought him into closer contact with students, but here again not in a way that was characteristic of university professors. His attitude toward students was characterized chiefly by his friendliness and readiness to help them. When a student really had a problem in which he was profoundly interested, even if it was a very simple one, Einstein was ready to devote any amount of time and effort to help him solve it. Also the incredible ease with which Einstein instituted even difficult scientific reflections and his almost equally unusual talent for comprehending rapidly and thoroughly what was said to him stood him in good stead in these consultations. As a result he had a good deal of time, which he lavishly placed at the disposal of his students.

When I came to Prague as his successor, Einstein's students told me with the greatest admiration and joy that immediately upon assuming his duties as professor there, he had said to them: "I shall always be able to receive you. If you have a problem, come to me with it. You will never disturb me, since I can interrupt my own work at any moment and resume it immediately as soon as the interruption is past."

This attitude must be judged in comparison with that of many professors, who tell their students that they are always occupied with their research and do not like to be disturbed because an interruption might possibly imperil the results of their intensive reflections.

Just as it is the pride of many people never to have any time, so it has been Einstein's always to have time. I recall a visit I once paid him on which we decided to visit the astrophysical observatory at Potsdam together. We agreed to meet on a certain bridge in Potsdam, but since I was a good deal of a stranger in Berlin, I said I could not promise to be there at the appointed time. "Oh," said Einstein, "that makes no difference; then I will wait on the bridge." I suggested that that might waste too much of his time. "Oh no," was the rejoinder, "the kind of work I do can be done anywhere. Why should I be less capable of reflecting about my problems on the Potsdam bridge than at home?"

And this was very characteristic of Einstein. His thoughts flowed like a constant stream. Every conversation that interrupted his thinking was like a small stone thrown in a mighty river, unable to influence its course.

There was yet another factor that brought Einstein into closer contact with his students. This was his need to clarify his ideas for himself by expressing them aloud and explaining them to others. Thus he often conversed with students about scientific problems and told them his new ideas. But Einstein did not really care whether the listener actually understood what was being explained or not; all that was necessary was that he should not appear too stupid or uninterested. Einstein once had an assistant who helped him with his administrative duties while at the same time completing his own studies in physics. Every day Einstein explained his new ideas to him and it was generally said that if this young man had had only a slight talent, he could have become a very great physicist — few students had ever received such good instruction. But while the student was an intelligent and industrious man and an ardent admirer of Einstein, he did not become a great physicist. The influence of the teacher is not so great as some people believe.

---

7. *Outbreak of the World War*

Before Einstein had been in Berlin a whole year, the World War broke out in August 1914. Great enthusiasm swept Germany, which to a large extent arose from the feeling that the individual could now merge with the greater whole — the German Empire — and stop living for himself, a feeling that for many people meant a great sense of relief.

This joy, however, could not be felt by anyone who had any comprehension of public opinion in the great Slavic centers of Austria. In Prague Einstein had witnessed the gradual evolution by which Austrian foreign policy had become an instrument for the attainment of German aims, and consequently Einstein could not share the enthusiasm of the crowd in Berlin. He was placed in a rather unpleasant psychological situation. His feelings were comparable to those of a person in the midst of a group that has been stimulated by good wine, but who has drunk nothing himself. He felt badly about it, because he represented for the others a sort of quiet reproach, which they resented. Fortunately, he had a good reason to base his reserve on. In coming to Berlin, he had retained his Swiss nationality, and his lack of enthusiasm as a neutral was not taken too much amiss.

I still remember very clearly the first visit that I paid Einstein during the war. When I was leaving, he said to me: "You have no idea how good it is to hear a voice from the outside world, and to be able to speak freely about everything."

Immediately upon the outbreak of the war there arose behind the actual battle front an "intellectual front," where the intellectuals of the hostile camps attacked each other and defended themselves with "intellectual weapons." The invasion of neutral Belgium by German troops had shocked the entire world, which still believed in the validity of "paper" treaties. Furthermore, the suffering of the Belgian people during the fighting and occupation was utilized to very sound advantage by Allied propaganda. The people of western Europe asked with astonishment: "How can the German people, whose music we love and whose science we admire, be capable of such unlawfulness and such atrocities?" Partly for propagandist reasons there was invented the story of the "two Germanys," the Germany of Goethe and the Germany of Bismarck.

The creation of this contrast was unpleasant for the German government, which demanded of the intellectuals that they publicly proclaim their solidarity with the German military and diplomatic conduct of the war. In the famous *Manifesto of the Ninety-two German Intellectuals,* ninety-two of the outstanding representatives of German art and science rejected the distinction between German culture and German militarism. The manifesto culminated in the assertion: "German culture and German militarism are identical." What from the German side was regarded as a disavowal of disunity in the life-struggle of the nation was considered the height of cynicism by the Allies.

As one might expect, Einstein did not sign the manifesto. But it illustrates what was expected at that time of every leading German artist and scientist. Anyone, like Einstein, who refused to concur was regarded by the great majority of his colleagues as a renegade who had deserted his people at a difficult time. Only his Swiss citizenship saved Einstein from being looked upon as a traitor in the struggle for the existence of the German people.

One can understand how difficult it would have been for Einstein to identify himself publicly with that very militarism to which he had had the greatest aversion since childhood.

---

8. *German Science in the War*

With the outbreak of the war, all of Einstein's colleagues became active in one way or another in war service. Physicists were employed in wireless telegraphy, in constructing submarine sound detectors, in predicting weather, and various other important scientific projects. Some served because they felt it to be their duty, others because such work was less unpleasant than service on the battle fronts. On the other hand, there were some who felt they should share the dangers and hardships of the soldiers in the trenches instead of working in a safe laboratory.

Walter Nernst, who has been mentioned several times already, performed valuable services in the investigation of poison gases. Fritz Haber, a close scientific friend of Einstein, developed a process for the manufacture of ammonia utilizing atmospheric nitrogen, a process of great significance since ammonia is a chemical necessary for the manufacture of artificial fertilizers and explosives and since Germany was unable to import natural ammonia compounds because of the English blockade. Haber was of Jewish origin, but he was strongly influenced by such Prussian ideas as high regard for military power and the subordination of personal feelings to this supreme value. For their service both Nernst and Haber received the rank of major in the German army. To Nernst this title was only a minor satisfaction to his vanity, and he did not esteem it very highly, but for Haber it was a source of great satisfaction and sincere pride. In the Treaty of Versailles both Nernst and Haber were listed by the Allies among the "war criminals" whom Germany was supposed to give up for trial before an international court. No serious demand for their surrender was ever made, however.

All this work which the scientists performed for the war effort was only natural at a time of national peril, no matter what their attitudes were to the government in power. But there was still another way in which they participated in the war: they engaged actively in the war on the "intellectual front." There began a battle of words and of propaganda by which the achievements of German scientists were stressed while those of the workers in the enemy countries were depreciated. A group of German physicists sent a circular to all their colleagues in which they urged them not to cite the works of English physicists, or to do so only where this was "unavoidable." They asserted that on the whole the work of Englishmen was on a much lower level and was frequently mentioned only because of an exaggerated admiration for foreigners, an attitude that should now be abandoned.

From a historical point of view it is not so much these humanly understandable attempts to exploit the war spirit for personal advantages that are of interest, but rather the apparently "scientific" attempts to prove that the entire structure of German physics differs from that of French or English physics. It was argued that for this reason one should adopt as little as possible from them, since otherwise the unity and purity of German science would be endangered and the minds of German students confused. For example, it was frequently asserted that German science is especially profound and thorough in contrast to the superficial character of French and Anglo-American science. French superficiality was attributed to the "shallow" rationalism that tries to comprehend everything by means of reason and ignores the mystery of nature; that of the Anglo-Saxons to the overemphasis on sensory experience, which believes only in facts and ignores philosophical implications.

Against this the French scientists, in so far as they participated in the war of words, asserted that the "thoroughness" of German science consists in a pedantic collection of unimportant facts, and its "philosophical" character in the production of a smoke screen that obscures the true relation between things. The Anglo-Saxon scientists preferred to point out that German science emphasizes "idealistic" principles so as to make it easier to excuse particularly inhuman acts; for if one must commit atrocities in order to carry out such principles, then they are "idealistically" justified.

These arguments soon made their appearance in the controversy over the relativity theory. By using arguments of this kind it could be attacked by the one party as a particularly "German" theory and by the opposing side as particularly "un-German." We shall see that in this way Einstein's theories, which at first sight appear far removed from any political utility, were soon drawn into the struggles of nations and parties.

9. *Life in Wartime*

During the war the newspapers in Berlin were filled with the battles and victories of the German army. The people were filled with joy and occupied themselves with discussions of such questions as which of the conquered territories should be kept by Germany after the war, whether Poland should actually be freed or become a German protectorate, and so forth. They counted the number of English merchant ships sunk by their submarines, and many of them kept lists of the amount of tonnage sent to the bottom of the sea. Every day they copied the figures from the newspapers and conscientiously added up the totals like a business man making up his annual accounts. To their astonishment they soon found that the total exceeded all the tonnage England ever had, and they began to wonder that there were still any English ships on the sea.

In private life, however, the pre-eminent interest of everybody was in obtaining food. Whoever managed a household had to be as cunning and ingenious as possible to get any of the food that occasionally appeared on the market; and to prepare it in a halfway palatable manner, since it was often of an unusual nature.

Einstein's health was often poor during the war, and he was happy to be connected with a family with whom he could eat home-cooked meals instead of having to depend on restaurant fare, whose cooking at this time was based on the hygienic instructions of the military authorities. Some of Einstein's well-to-do relatives had previously looked upon him as the black sheep of the family. His running away from the gymnasium in Munich, his devotion to studies that could not bring him a good income, and his marriage to a woman completely outside their circle had not met their approval. It was therefore with great astonishment that they had heard of his growing fame. When Einstein was called to Berlin and made a member of the Royal Prussian Academy, they felt honored to have him at their homes and to be mentioned as his relatives. Einstein accepted this situation good-humoredly.

In his uncle's house Einstein again met his cousin Elsa, with whom he had been friends as children in Munich. She was now a widow with two daughters, a woman of friendly, maternal temperament, fond of amusing conversation, and interested in creating a pleasant home and preparing the scanty wartime meals as best she could. Einstein often went to their house, and found a new family life there.

Frau Elsa could not study the works of great physicists with him as Mileva Maritsch had done at Zurich. She had a happy outlook on life, and not the harsh, self-denying nature of the Slavic student. Regarding Einstein as a physicist, she knew only that he had now become a famous man whom the outstanding scientists of the Prussian Academy, the University of Berlin, and foreign countries recognized as their equal and often as their superior. To have such a relative and friend was a source of pride and joy to her and she wanted to relieve him of the cares of daily life. Einstein, who valued friendliness, often made himself useful in her house by practicing "applied physics."

When I visited Berlin on one occasion during the war, Einstein invited me to his uncle's house for dinner. I declined at first, saying: "Right now when everything is so scarce no one likes to have an unexpected guest." Thereupon Einstein replied in his sincere manner, which sounded like the simplicity of a child but which could equally well be regarded as acid criticism: "You need have no scruples. On the contrary, my uncle has more food than the per capita average of the population. If you eat at his table you are serving the cause of social justice." There I met his cousin Elsa for the first time. She said to me half playfully, half in earnest: "I know very well what a talented physicist our Albertle is. In these times we have to buy food in all kinds of cans which no one knows how to open. Often they are of unfamiliar, foreign make, rusted, bent, and without the key necessary to open them. But there hasn't been a single one yet that our Albertle has not been able to open."

While the war was still going on, Einstein married his cousin Elsa. He, who had always had something of the bohemian in him, began to lead a middle-class life. Or, to put it more exactly, Einstein began to live in a household such as was typical of a well-to-do Berlin family. He lived in a spacious apartment in the so-called "Bavarian quarter." This section had nothing Bavarian about it except that the streets were generally named after Bavarian cities. He lived in the midst of beautiful furniture, carpets, and pictures; his meals were prepared and eaten at regular times. Guests were invited. But when one entered this home, one found that Einstein still remained a "foreigner" in such a surrounding — a bohemian guest in a middle-class home.

Elsa Einstein had many of the characteristics of the people of her native Swabia. She valued greatly what was known in Germany and especially in Swabia as *"Gemütlichkeit."* It is no wonder that she was very happy when she saw the esteem and admiration in which her husband was held and which she shared as his wife. Nevertheless, there were always two sides to the job of being the wife of a famous man. The people about her were always inclined to look very critically at her and, as a compensation for the respect that they reluctantly paid her husband, to unload upon her all the reproaches they would have liked to bring against him.

When Elsa Einstein was discussed in professional circles in Berlin, one could hear all sorts of criticism of this nature. The most harmless was probably the assertion that her intellectual capacities hardly fitted her to be Einstein's wife. But if Einstein had followed this criticism, what woman could he have married? The question was, rather, could she create tolerable living conditions for Einstein in which he could carry on his work? And in considerable measure she did so. There is no ideal solution to this problem, and since Einstein believed less than most men in the possibility of an ideal solution, he did not feel hurt when his wife did not completely represent this ideal.

Some professors complained that because of her it was difficult for physicists to gain access to Einstein. She preferred, they claimed, to have Einstein meet writers, artists, or politicians, because she understood these people better and considered them more valuable. Einstein, however, was certainly not the man to be easily influenced in the choice of his company. He himself liked to mingle with all kinds of people and did not restrict himself to professional circles. It may sometimes have happened that a visitor whom Einstein did not wish to see put the blame on Einstein's wife because he did not want to admit to himself that his company was not so interesting for Einstein as he himself thought it ought to be.

Others complained that Mrs. Einstein placed too much value on the external symbols of fame and did not really know how to value her husband's inner greatness. It is obvious, however, that the wife of a great man can understand most easily the effect of his activities on public opinion, and that this will consequently interest her more than anything else.

Any woman in Elsa Einstein's position would probably have acted more or less as she did. The only difference was that the public is rarely so much interested in the life of a scientist as it was in Einstein's. On this account his wife was blamed for various things that are actually common occurrences. The married life of a great man has always been a difficult problem, no matter how he or his wife is constituted. Nietzsche once said: "A married philosopher, is to put it bluntly, a ridiculous figure."

Einstein was protected against various difficulties by the circumstance that he always kept a certain part of his inner self from any contact with others, and that he had no desire to share his inner life completely with anyone. He was very much aware that every happiness has its shadows, and accepted this fact without protest.

When in 1932 some women's clubs opposed Einstein's entry into the United States because in their opinion he spread subversive doctrines, e.g., pacifism, Einstein remarked jokingly to a representative of the Associated Press: "Why should one admit a man who is so vulgar as to oppose every war except the inevitable one with his own wife?"

And on another occasion he made a remark based on many years of experience: "When women are in their homes, they are attached to their furniture. They run around it all day long and are always fussing with it. But when I am with a woman on a journey, I am the only piece of furniture that she has available, and she cannot refrain from moving around me all day long and improving something about me."

This lack of any illusion about the possibility of happiness in life has saved Einstein from the mistake made by many a husband who looks upon all the defects that are characteristic of life itself as defects in his wife and in consequence plays the stern judge with her instead of remembering her good qualities and accepting her bad ones as a necessity of nature.

During this period Einstein's first wife and his two sons lived in Switzerland. This circumstance caused Einstein a great deal of financial worry because of the great difficulty in transferring money from Germany to Switzerland and the rate of exchange, which became more and more unfavorable as the war progressed. But since her student days Mileva Maritsch was so attached to her life in Switzerland that on no account would she live in Germany.

VI

THE GENERAL THEORY OF RELATIVITY

1. *New Theory of Gravitation*

The war and the psychological conditions produced by it in the world of science did not prevent Einstein from devoting himself with the greatest intensity to improving his theory of gravitation. Working along the line of his ideas that he had found in Prague and Zurich, he succeeded in 1916 in developing a completely independent, logically unified theory of gravitation. Einstein's conception differed fundamentally from that of Newton, and a real understanding of his theory requires a wide knowledge of mathematical methods. Without using any mathematical formulæ, I shall here attempt to present the fundamental ideas in so far as they are necessary for our understanding of Einstein's personality and the influence of his theory on his period and environment.

The great difficulty involved in explaining Einstein's new theory lies in the fact that it does not arise from any slight modification of Newtonian mechanics. It bursts asunder the entire framework within which Newton attempted to comprehend all phenomena of motion. The familiar concepts of "force," "acceleration," "absolute space," and so on have no place in Einstein's theory. Even to the average physicist the principles composing Newtonian mechanics seem either to be proved by experience or by reasoning, and it is hardly possible for him to comprehend any change in a structure that he has come to regard as immutable. This is an illusion that must be destroyed in order to be able to understand Einstein's theory.

According to Newton's law of inertia, a body not acted on by any force moves in a straight line with constant velocity. This is true no matter what is the mass or other physical properties of the body involved. Hence it may be stated that its motion can be described "geometrically." On the other hand, if any force acts on the body, then, according to Newton's law of force, it experiences acceleration inversely proportional to its mass. Consequently, particles with different masses perform different paths under the action of the same force. Motion under force can only be described by using a non-geometrical term *mass*.

We have seen in Section 8 of Chapter IV, however, that in his gravitational theory of 1911 Einstein had noted that the force of gravity has the unique property that its influence is independent of the mass of the body on which it acts. And as a consequence he had concluded that the presence of a gravitational field of force cannot be distinguished from the result of accelerated motion of the laboratory. This means that not only motion under no force, but also motion under gravitational force alone can be described purely geometrically, if these forces are parallel and of equal magnitude in the whole region considered.

With this foundation, the problem that now faced Einstein was this: What is the geometrical form of the path which a body in a gravitational field describes relative to any laboratory?

Einstein's solution of this problem is based on a concept that the laws of geometry in a space where there exists gravitational field are different from those in a space which is "free of forces" in the old sense. This was an idea so novel that the physicists and mathematicians used to nineteenth-century physics were bewildered by it. In order to understand what Einstein meant, we must go back to the positivistic conception of science, and in particular to the ideas of Henri Poincaré described in Section 9 of Chapter II. According to this view, the truth of mathematical propositions concerning points, straight lines, and so forth can only be verified in our world of experience when these mathematical notions are defined in terms of physical operations. We must give what P. W. Bridgman calls the "operational definitions" to the geometrical terms. For example, we must define "straight lines" in terms of certain steel rods prepared according to a specified method, and if we make a triangle with these rods we can verify by actual measurement on this triangle whether the angles add up to two right angles or not.

By means of other experiments we can then investigate whether these rods actually have all the properties that geometry postulates about "straight lines." For instance, we can measure whether such a rod is really the shortest line connecting two points. Of course in order to be able to carry out this measurement we must also describe a physical operation of measuring the length of a curved line. It may be found that when a triangle is formed by joining these points by lines that form the shortest distances between these points, the sum of the angles of this triangle does *not* equal two right angles. We are then faced with a dilemma. If we say that the lines forming this triangle are straight lines, we retain the property of the straight line to be the shortest distance between any two points, but then the theorem of the sum of angles is no longer valid. On the other hand, if we want the theorem to be valid, the property to be the shortest distance has to be rejected. We are free to decide which property we retain for the lines we call "straight," but we cannot have them both as in Euclidean geometry.

Einstein's fundamental assumption can now be re-expressed in this form: In a space where masses that exert gravitational forces are present, Euclidean geometry ceases to be valid. In this theory curves which are the shortest distances between any two points have special significance, and the angles of a triangle formed by these lines do not add up to two right angles where a gravitational field exists.

This distinction between Euclidean space and the "curved" space of Einstein can be illustrated by considering a similar distinction between a plane surface and a curved surface. For all triangles on a plane surface, all of Euclid's theorems hold true; but what happens for triangles on a curved surface? Take for example the surface of the earth. If we are restricted to only those points which actually lie on the surface and cannot consider any point lying above or below it, there are no "straight lines" in the usual sense. But the curves which form the shortest distance between two points on the earth's surface are important in navigation and geodesy; they are called geodesic lines. For the surface of the sphere, the geodesic lines are arcs of great circles, and consequently all the meridians defining longitude and the equator are geodesic. If we consider a triangle formed by the North Pole and two points on the equator it is bounded by geodesic lines. The equator cuts all the meridians perpendicularly so that the two angles at the base of the triangle are both right angles, and hence the *sum of the angles is greater than two right angles* by just the value of the polar angle. A similar situation always holds for any curved surface, and, conversely, if the sum of the angles of a triangle formed by geodesic lines on a surface does not equal exactly two right angles, then the surface is curved.

This notion of curvature of a surface is extended to space. Geodesic lines are defined as curves forming the shortest distances between any two points in space, and the space is called "curved" if the angles of a triangle formed by three geodesic lines do not add up to two right angles. According to Einstein's theory, the presence of material bodies produces certain curvatures in space, and the path of a particle moving in a gravitational field is determined by this curvature of space. Einstein found that such paths can be described most simply by considering the geometry of this curved space rather than by ascribing its deviation from a straight line to the existence of forces as Newton had done. Furthermore, Einstein found that not only the paths of material particles, but also those of light rays in a gravitational field can be described simply in terms of geodesic lines in this curved space; and, conversely, that the curvature of space can be inferred from observations on the path of moving bodies and light rays.

We shall see later that many people, even some physicists, considered it absurd to say that any conclusion about the curvature of space can be drawn from the form of light rays. Some even considered it completely nonsensical to say that a space is "curved." To them, a surface or a line may be *curved in space,* but to say that space itself is "curved" seemed preposterous and absurd. This opinion, however, is based on ignorance of the geometrical mode of expression. As we have seen above, a "curved space" simply means a space in which the sum of the angles of a triangle formed by geodesic lines does not equal two right angles, and this terminology is used because of the analogous distinction between flat and curved surfaces. It is futile to try to picture what a curved space "looks" like, except by describing the measurement of triangles.

2. *Role of Four-Dimensional Space*

If we wish to describe the motion of a certain particle completely, it is not sufficient to give the shape of its trajectory, but it is necessary to add how the position of the particle on this trajectory varies with the time. For instance, to say that the motion of a particle uninfluenced by any force in the Newtonian sense is rectilinear is not complete; we must add that its motion takes place with constant velocity.

The complete motion can, however, be presented in a geometrical form by adding a dimension to the number necessary to describe the trajectory. For example, in the simplest case of a rectilinear motion, the trajectory is a straight line, and the position of the particle describing it can be specified by giving the distance that the particle is from a certain definite point on the straight line. We now take a sheet of paper and plot these distances along one direction, and for each point plot in a direction perpendicular to the distance the time corresponding to each position. Then the curve drawn through these points gives the complete geometrical presentation of the motion. If the motion takes place with constant velocity as well as being rectilinear, the curve will be a straight line. Thus motion along a straight line, or one-dimensional motion, to express it technically, can be represented *completely* on a plane — that is, in two-dimensional space. Now, the space of our experience has three dimensions; to specify the position of a ball in a room, we must give three numbers, the distances from the two walls and its height above the floor. Hence we need three dimensions to describe the trajectory of a general motion, and four dimensions to give a *complete* presentation of the motion. The motion of a particle is specified completely by a curve in a four-dimensional space.

This notion of four-dimensional space, simple as it is, has given rise to a great deal of confusion and misunderstanding. Some writers have maintained that these curves in four-dimensional space are "only aids for mathematical presentation" and "do not really exist." The statement "do not really exist," however, is a pure truism, since the statement "really existing" is used in daily life to describe only directly observable objects in our three-dimensional space. In contrast to this, many authors, especially philosophers and philosophically tinged physicists, have taken the point of view that *only the events in four-dimensional space are real,* and a representation in three-dimensional space is only a subjective picture of reality. We can readily see that such a position is equally justified except that the word "real" is used in a different sense. To clear up this disagreement we have to use a little *semantics.*

In his special theory of relativity developed at Bern, Einstein had shown that when mechanical and optical phenomena are described by means of clocks and measuring rods, the description depends on the motion of the laboratory in which these instruments are used. And he had been able to state the mathematical relations that correlate the various descriptions of the same physical event. In 1908 Hermann Minkowski, Einstein's former professor of mathematics at Zurich, showed that this relationship between different descriptions of the same phenomena can be represented mathematically in a very simple manner. He pointed out that these different descriptions of a motion represented by a curve in four-dimensional space are mathematically what are known as "projections of this four-dimensional curve on different three-dimensional spaces." Minkowski therefore took the view that only the four-dimensional curve "really" exists, and the different descriptions are merely different pictures of the same reality. This concept is analogous to saying that a fixed object in three-dimensional space, say a house, "really exists," but that photographs of this house taken from various directions — that is, two-dimensional projections of the three-dimensional house — never represent reality itself, but only descriptions of it from different points of view.

Obviously the word "real" is not used in the same sense here as when we say that only the three-dimensional body is "real" and that the four-dimensional presentation is simply an invented mathematical schema. In Minkowski's speech "real" means the "simplest theoretical presentation of our experiences," while in the other sense it means "our experience expressed as directly as possible in ordinary, everyday language."

Einstein's theory of gravitation started out from this representation of motion as a curve in four-dimensional space. Motion, if neither gravity nor any other force is acting, is represented by the simplest curve, the straight line, in flat, four-dimensional space. If only gravity but no other force is acting, Einstein assumed that the space becomes curved, but the motion is still represented by the simplest curve in such a space. Since there are no straight lines in curved space, he took for the simplest curve in space the curve with the shortest length between any two points — that is, the geodesic line. Hence motion of a particle under gravity is represented by a geodesic curve in four-dimensional curved space, and this curvature of space is determined by the distribution of matter which produces the gravitational field.

Thus Einstein's general theory of relativity consists of two groups of laws:

First: The field laws which state how the masses present produce the curvature in space.

Second: The laws of motion both for material particles and for light rays which state how the geodesic lines can be found for a space whose curvature is known.

This new theory of Einstein was a fulfillment of the program of Ernst Mach. From the material bodies present in space it enables one to calculate the curvature of space, and from this the motion of bodies. According to Einstein, the inertia of bodies is not due, as Newton has assumed, to their efforts to maintain their direction of motion in absolute space, but rather to the influence of the masses about them — the fixed stars, as Mach had suggested.

---

3. *Einstein Suggests Experimental Tests of His Theory*

Einstein's new theory, which so boldly and fundamentally changed the tested and successful Newtonian theory, was originally based on arguments of logical simplicity and generality. The question naturally arose whether new phenomena could be deduced from this theory which differed from those derived from the old, and which could be used as experimental tests between the two theories. Otherwise Einstein's theory remained only a mathematical-philosophical construction, which provided a certain degree of mental stimulation and pleasure but contributed nothing about physical reality. Einstein himself always recognized a new theory only if it uncovered a new field of the physical world.

Einstein showed mathematically that in "weak" gravitational fields his theory predicted the same results as Newton's. Here the curvature of our three-dimensional space is negligible, and the only difference comes from the new mathematical approach in the addition of the fourth dimension. The calculation of motion — for example, that of the earth around the sun — gives exactly the same result as that obtained from Newton's law of force and his theory of gravitation. It is only when the velocity of a body is comparable to that of light that any difference between the two theories can be detected.

In order to find phenomena where spatial curvature plays possibly a role, Einstein searched among the observations of celestial bodies for motions that were inconsistent with the predictions of Newtonian mechanics. He found one case. It had long been known that Mercury, a planet close to the sun and thus strongly exposed to its gravitational field, did not move exactly as predicted by Newton's theory. According to the old theory, all planets should perform elliptical orbits whose position in space are fixed in relation to the stars, but, it had been observed that the elliptical orbit of Mercury rotates around the sun at the very small rate of 43.5 seconds of an arc per century. This discrepancy had never been given a satisfactory explanation. When Einstein calculated the motion of Mercury according to his theory, he found that the orbit should actually rotate as observed. From the very beginning this achievement has been a strong argument in favor of Einstein's theory.

The effect of the curvature of space on the path of light rays is more impressive. While still at Prague Einstein had pointed out the possibility of bending rays of light as they passed close to the surface of the sun. He had calculated, on the basis of Newton's law of force and his own theory of gravitation of 1911, that the deflection should be 0.87 seconds of an arc. According to his new theory of curved space, Einstein found the deflection is 1.75 seconds, actually twice as great as his former result.

The third prediction that Einstein made was on the change in wave length of light emitted by a star. His calculation showed that light, in leaving the star where it is emitted, has to pass through its gravitational field, and this passage shifts the wave length toward the red. Even for the sun the effect turned out to be hardly observable, but in the case of the very dense companion star to Sirius it seemed to be of observable magnitude.

It is important to note that of these three phenomena predicted by the theory, only one of them, the motion of Mercury, was actually known at the time Einstein developed his theory. The other two were entirely new phenomena, which had never been observed or even suspected. Both of these predictions received unqualified verification some years later, and so gave conclusive evidence of the correctness of the theory. It is very remarkable and a great tribute to Einstein that he was able to develop a theory which started from few fundamental principles and using the criterion of logical simplicity and generality led to amazing results.

4. *Cosmological Problems*

Even before his new theory had been completely understood by the great majority of physicists, it was already evident to Einstein that it was unable to give a correct presentation of the universe as a whole.

During the nineteenth century the commonest conception of

the universe was that there are groups of material bodies like our Milky Way, and outside this region is "empty" space, which extends infinitely far. This view had, however, already aroused doubts among some scientists around the end of the century. For in this case the stars would behave like a cloud of vapor and there was nothing to prevent them from dispersing into the surrounding empty space. Since infinite time and space are available, the whole universe would eventually become completely empty.

From the standpoint of Einstein's theory, this conception of the material universe as an island in empty space had additional difficulty. This is due to the equivalence principle, by which gravitational and inertial masses are considered identical. It will be remembered that Ernst Mach first pointed out as a defect of Newtonian mechanics that in it inertial motion, rectilinear motion in empty space, is a process uninfluenced by the presence of other masses. Mach proposed instead the assumption that the effect of inertia is due to motion relative to the fixed stars. Einstein had introduced this idea in his theory as "Mach's postulate" when he assumed that gravitational field, and consequently inertial effects, are determined by the distribution of matter. If the material bodies formed an island in empty space, then, according to Einstein, only a finite part of space would be "curved." This region, however, would be surrounded by a "flat" space extending to infinity in all directions. In this flat space, bodies not acted on by any force would move in straight lines in accordance with Newton's law of inertia, and the inertial force would not be determined by the distribution of matter. For this reason, the idea of curved space being enclosed in an infinite flat space is inconsistent with Mach's postulate.

The next possible assumption then was that matter does not form an island, but rather that all of space is filled more or less densely with matter. However, if we further assume that all these masses act upon each other according to Newton's law, then we again run into a difficulty. For matter at large distances exerts individually small effects, but the total amount of matter at large distances increases in such a way that there is an infinite amount of matter at infinity which exerts an infinitely strong force. Observations show that stars are not acted on by such forces, for in this case they would reach high velocities, while all actually observed velocities of stars are small in comparison with the speed of light.

Einstein cleared up this difficulty by pointing out that in his

theory of curved space uniform distribution of matter does not
necessarily mean that there is an infinite amount of matter.
There is the possibility that, owing to the curvature, space does
not extend to infinity. This does not mean, however, that there
are boundaries in space beyond which is nothing, not even empty
space. The situation may perhaps be illustrated by the same example with which I explained the curvature of space. The surface of the earth is a two-dimensional curved surface which has
finite area but has no boundaries. Certain objects, say cities, may
be distributed more or less uniformly on its surface, but the total
number of cities is finite. Furthermore, if one travels in a given
direction along any geodesic (a great circle in this case), one
returns to the original point of departure. In the same way the
space of our experiences may be curved in such a way that it
is finite but unbounded. It becomes meaningful to ask how
much matter is contained in the universe, what is the "radius
of curvature" of our space, and consequently what is the average density of matter in space.

There is still another possibility, however. Matter may fill "infinite" space with approximate uniformity, but the whole universe may not be at rest, but expanding, so that the density of
matter is decreasing. At present it is not yet possible to say with
certainty which of the two hypotheses concerning the distribution of matter is correct. Later on, Einstein envisaged also the
possibility that space might be "curved" without the presence of
masses, contrary to Mach's original assumption.

At any rate, the view that matter does *not* form an island in
infinite empty space is supported by modern astronomy. The
researches of Harlow Shapley and his collaborators have shown
that space, as far as can be seen with present telescopes, seems
to be similar everywhere to the region of our Milky Way. Thus
it is plausible to assume with Einstein that on the average the
entire universe is uniformly filled with matter. Also by count-
ing the number of stars and measuring their distances from us,
Shapley has been able to obtain a rough value for the average
density of matter in the universe. Furthermore, from observations of the velocity of recession of the distant nebulæ and Einstein's law of motion it has been possible to calculate such quantities as the radius of curvature and the volume of space, and
the total amount of matter in it.

5. *Expeditions to Test Einstein's Theory*

For the mathematician, Einstein's new conception of gravitation was characterized by beauty and logical simplicity. For the observational astronomer there still remained the disquieting doubt that all this might be mere fantasy. Newton's theory had served them well and it would require more than mathematical elegance to change their views. According to the astronomers, a solar eclipse was needed for the test.

New theories — to use a comparison that Einstein likes to employ — are comparable to beautiful dresses, which when displayed in a dressmaker's salon attracts every feminine eye. A celebrated beauty orders this dress, but will it fit her? Will it add to or detract from her beauty? Not until she has worn it in the full glare of lights can she tell. Einstein's theory was a kind of unworn dress that had been in a shop window. The solar eclipse was the first affair at which it was to be worn.

While the war was still in progress, Einstein's papers on the general theory of relativity became known in England. The abstract discussion could be followed only with difficulty, and the new conceptions about motion in the universe could not yet be appreciated in all their logical implications. But their boldness was already admired. For the first time a well-founded proposal had been advanced to change the laws of the universe set up by Isaac Newton, England's pride.

For the English, with their tendency toward experimental verification, one thing was clear. A number of definite experiments had been pointed out to the observer of nature whose results could give decisive evaluation to the merits of the theory. And among these it was pre-eminently Einstein's prediction on the shift in the position of the stellar images during a total solar eclipse that made it possible to test his two theories, the Prague theory of 1911 and the Berlin theory of 1916. As early as March 1917 the Astronomer Royal had pointed out that on March 29, 1919 a total solar eclipse would take place that would offer unusually favorable conditions for testing Einstein's theories, since the darkened sun would be situated in the midst of a group of particularly bright stars, the Hyades.

Although at that time no one knew whether it would be possible to send expeditions to those regions of the earth where the observation of the total eclipse would be possible, the Royal Society and the Royal Astronomical Society of London appointed a committee to make preparations for an expedition. When the armistice was signed on November 11, 1918, the committee immediately set to work and announced the detailed plans for the expedition on March 27. The committee was headed by Sir Arthur Eddington, one of the few astronomers who were able at that time to delve deeply into the theoretical foundations of Einstein's theories. Eddington, moreover, was a Quaker who had always attached great importance to the maintenance of a friendly feeling between the people of "enemy" nations, and both during and after the war he did not join in the customary feeling of hate for the enemy. He also regarded all new theories about the universe as a means of strengthening religious feeling and of directing the attention of people away from individual and national egoism.

When the sun is eclipsed by the moon, there is only a certain zone on the earth's surface where the entire solar disk is darkened. Since there is the chance that the weather may be poor during the few minutes of darkness and thwart all plans of observation, the Royal Society sent two expeditions to widely separated points within the zone of total eclipse. One set out for Sobral in northern Brazil, while the second sailed for the isle of Principe in the Gulf of Guinea, West Africa. Eddington was in personal charge of the second group.

When the expedition arrived in Brazil, it aroused not a little astonishment and something of a sensation. The war with Germany was hardly over, and the newspapers were still full of propaganda and counter-propaganda. These had not spared scientific activities, but yet here was a costly expedition coming from England to test the theories of a German scientist. A newspaper in Pará, Brazil, wrote: "Instead of trying to establish a German theory, the members of the expedition, who are well acquainted with the heavens, should rather try to obtain rain for the country, which has suffered from a long drought." The expedition was really in luck, since several days after its arrival it began to rain in Sobral. The savants had justified the public's confidence in science.

I shall not describe the observations made in Brazil, but merely those made by the group on the isle of Principe. The astronomers arrived a month before the date of the eclipse in order to set up their instruments and to make the necessary preparations. And then came the few minutes of total eclipse, with the disquieting uncertainty whether it would be possible to photo-

graph the stars in the neighborhood of the darkened sun, or
the clouds would hide the stars and nullify the months of preparation. Sir Arthur Eddington gave the following description
of these moments:

"On the day of the eclipse the weather was unfavourable. When
totality began, the dark disc of the moon surrounded by the corona
was visible through cloud, much as the moon often appears through
cloud on a night when no stars can be seen. There was nothing for it
but to carry out the arranged programme and hope for the best. One
observer was occupied changing the plates in rapid succession, whilst
the other gave the exposures of the required length with a screen
held in front of the object-glass to avoid shaking the telescope in any
way.

> For in and out, above, about, below
> 'Tis nothing but a Magic *Shadow*-show
> Played in a Box whose candle is the Sun
> Round which we Phantom Figures come and go.

"Our shadow box takes up all our attention. There is a marvellous
spectacle above and as the photographs afterwards revealed, a wonderful prominence flame is poised a hundred thousand miles above
the surface of the sun. We have no time to snatch a glance at it. We
are conscious only of the weird half-light of the landscape and the
hush of nature, broken by the calls of the observers and the beat of
the metronome ticking out the 302 seconds of totality.

"Sixteen photographs were obtained, with exposures ranging from
2 to 20 seconds. The earlier photographs showed no stars . . . but
apparently the cloud lightened somewhat towards the end of totality,
and a few images appeared on the later plates. In many cases one or
the other of the most essential stars was missing through cloud, and
no use could be made of them; but one plate was found showing
fairly good images of five stars, which were suitable for a determination."

Tense with excitement, Eddington and his collaborators compared the best of the pictures that they had obtained with photographs of the same stars taken in London, where they were far
removed from the sun and therefore not exposed to its direct
gravitational effect. There actually was a shift of the stellar images away from the sun corresponding to a deflection of the
light rays approximately as large as that expected on the basis
of Einstein's new theory of 1916 (fig. 3 and 4).

It was quite a few months, however, before the expeditions
had returned to England and the photographic plates were carefully measured in the laboratory, taking into consideration all
possible errors. These errors were what actually worried the ex-

perts. Around them revolved the discussions in astronomical circles, while the lay public was interested, and could only be interested, in the question whether the observations had demonstrated the "weight of light" or the "curvature of space." The latter was even more exciting since hardly anyone could imagine anything very definite under the phrase "curvature of space."

6. *Confirmation of the Theory*

On November 7, 1919 London was preparing to observe the first anniversary of the armistice. The headlines in the London *Times* were: "The Glorious Dead. Armistice Observance. All Trains in the Country Stop." On the same day, however, the *Times* also contained another headline: "Revolution in Science. Newtonian Ideas Overthrown." It referred to the session of the Royal Society on November 6, at which the results of the solar-eclipse expedition were officially announced.

The Royal Society and the Royal Astronomical Society of London had convened a combined session for November 6 to make the formal announcement that the expeditions that had been dispatched by these societies to Brazil and West Africa to observe the total solar eclipse had from their observations reached the conclusion that the rays of light are deflected in the sun's gravitational field and with just the amount predicted by Einstein's new theory of gravitation. This remarkable agreement between a creation of the human mind and the astronomical observations gave the session a wonderful and exciting atmosphere. We have an eyewitness account of this meeting by one of the most highly regarded philosophers of our time, Alfred North Whitehead. As a mathematician, logician, philosopher, and a man endowed with a fine historical and religious sense, he was better suited to experience the uniqueness of this hour than most scientists.

"It was my good fortune," said Whitehead, "to be present at the meeting of the Royal Society in London when the Astronomer Royal for England announced that the photographic plates of the famous eclipse, as measured by his colleagues in Greenwich Observatory, had verified the prediction of Einstein that rays of light are bent as they pass in the neighbourhood of the sun. The whole atmosphere of tense interest was exactly that of the Greek drama. We were the chorus commentating on the decree of destiny as disclosed in the development of a supreme incident. There was dramatic quality in the very staging — the traditional ceremonial, and in the background the picture of Newton to remind us that the greatest of scientific generalizations was now, after more than two centuries, to receive its first modification. Nor was the personal interest wanting; a great adventure in thought had at length come safe to shore.

"The essence of dramatic tragedy is not unhappiness. It resides in the remorseless working of things. . . . This remorseless inevitableness is what pervades scientific thought. The laws of physics are the decrees of fate."

At this time the president of the Royal Society was Sir J. J. Thomson, himself a great research physicist. He opened the session with an address in which he celebrated Einstein's theory as "one of the greatest achievements in the history of human thought." Continuing, he said: "It is not the discovery of an outlying island but of a whole continent of new scientific ideas. It is the greatest discovery in connection with gravitation since Newton enunciated his principles."

Then the Astronomer Royal reported in a few words that the observations of the two expeditions gave the value 1.64 seconds of an arc for the deflection of light, as compared with the value 1.75 seconds predicted by Einstein. "It is concluded," he announced briefly and dispassionately, "that the sun's gravitational field gives the deflection predicted by Einstein's generalized theory of relativity."

Sir Oliver Lodge, the famous physicist, who is widely known as an exponent of extra-sensory perception and other "parapsychological" phenomena, was always a convinced adherent of the existence of an "ether" that filled all space, and therefore hoped that the observations would decide against Einstein's theory. Nevertheless, after the session he said: "It was a dramatic triumph."

The scientists of the Royal Society were now ready to recognize that a direct observation of nature had corroborated the theory of the "curvature of space" and the invalidity of Euclidean geometry in gravitational field. Nevertheless, it was ominous of coming developments that during the formal session the president of the Royal Society himself said: "I have to confess that no one has yet succeeded in stating in clear language what the theory of Einstein really is." He persisted in his assertion that many scientists were themselves forced to admit their inability to express simply the actual meaning of Einstein's theory. It really meant that they were unable to grasp the meaning of

the theory itself; all they could understand were its consequences within their special field. This situation subsequently
contributed a good deal to the confusion of the lay public regarding Einstein's theory.

7. *Attitude of the Public*

The significance of the new theory was soon appreciated by men who were themselves creatively active in the development of science, but many of the so-called "educated"
people were annoyed that the traditional knowledge acquired
with great effort in the schools had been overthrown. Since such
people were themselves convinced of their lack of understanding of astronomy, mathematics, and physics, they attacked the
new theory in the fields of philosophy and politics, in which
they felt themselves qualified.

Thus an editorial writer in a reputable American newspaper
wrote of the session of the Royal Society described above: "These
gentlemen may be great astronomers, but they are sad logicians.
Critical laymen have already objected that scientists who proclaim that space comes to an end somewhere are under some
obligation to tell what lies behind it."

We recall that the statement: "Space is finite" has nothing to
do with an "end" of space. It means rather that light rays traveling through the world space return along a closed curve to their
origin. The editorial writers of daily newspapers like to represent the standpoint of the "man in the street," who is more often
influenced by a medieval philosophical tradition than by the
progress of science.

The editorial continues:

"This fails to explain why our astronomers appear to think that
logic and ontology depend on the shifting views of astronomers.
Speculative thought was highly advanced long before astronomy. A
sense of proportion ought to be useful to mathematicians and physicists, but it is to be feared that British astronomers have regarded
their own field as of somewhat greater consequence than it really is."

The same tendency to play off common sense — that is, in
this case, the knowledge acquired in elementary schools —
against the progress of science is also evident in another editorial
that appeared in the same reputable paper about this time:

"It would take the president of at least two Royal Societies to give plausibility or even thinkability to the declaration that as light has weight space has limits. It just doesn't, by definition, and that's the end of that — for common folks, however it may be for higher mathematicians."

Since in the opinion of the man in the street the two Royal Societies were affected by delusions which made them incapable of understanding things that were clear to anyone with an average school education, he began to inquire why such a thing had happened. An explanation was soon found, which was very illuminating for the man in the street.

One week after the famous London meeting, a professor of celestial mechanics at Columbia University, New York, wrote:

"For some years past the entire world has been in a state of unrest, mental as well as physical. It may well be that the war, the Bolshevist uprising, are the visible objects of some deep mental disturbance. This unrest is evidenced by the desire to throw aside the well-tested methods of government in favor of radical and untried experiments. This same spirit of unrest has invaded science. There are many who would have us throw aside the well-tested theories upon which have been built the entire structure of modern scientific and mechanical development in favor of methodological speculation and phantastic dreams about the Universe."

The writer then pointed out that the situation was analogous to the period of the French Revolution, when as a result of similar revolutionary mental diseases doubts were expressed concerning the Newtonian theory, though these objections later proved to be incorrect.

While various individuals were vexed by these innovations which disturbed their pride in their education, others received the matter in a more friendly manner. Einstein's predictions of the stellar shifts had shown, so these men thought, that physical phenomena could be predicted by means of pure thought, by pure mathematical speculation about the geometry of universal space. The view of the "wicked" empiricists and materialists that all science rests on experience, a view that caused so many conflicts with religion and ethics, had now been dropped by science itself. In an editorial dealing with the session of the Royal Society the London *Times* said: "Observational science has in fact led back to the purest subjective idealism." And "idealism" for the educated Englishman who received his education from his school, his church, and the *Times* was the diametrical opposite of "materialistic" Bolshevism.

The psychological situation in Europe at this time increased the interest of the general public in Einstein's theory. English newspapers tried to efface every connection between Germany and the man whom they were honoring. Einstein himself was averse to any tactics of this kind, not because he placed any value in being regarded as a representative of German science, but because he hated every manifestation of narrow-minded nationalism. He also believed that he could advance the cause of international conciliation if he utilized his fame for this purpose. When the *Times* requested him to describe the results of his theory for the London public, he did so on November 28 and used this opportunity to express his opinion in a friendly, humorous way. He wrote:

> "The description of me and my circumstances in the *Times* shows an amusing flare of imagination on the part of the writer. By an application of the theory of relativity to the taste of the reader, today in Germany I am called a German man of science and in England I am represented as a Swiss Jew. If I come to be regarded as a *'bête noire'* the description will be reversed, and I shall become a Swiss Jew for the German and a German for the English."

At that time Einstein did not anticipate how soon this joke would come true. The editor of the *Times* was slightly annoyed by the characterization of the way in which regard was taken of the prejudice of the middle-class British, and likewise answered in a semi-humorous vein: "We concede him his little joke. But we note that in accordance with the general tenor of his theory Dr. Einstein does not supply any absolute description of himself." The *Times* was also somewhat uneasy over the fact that Einstein did not have any feeling of belonging completely to a definite nation or race.

In Germany itself the news of the events in London acted like a spark that caused the explosion of pent-up emotion. It was a double satisfaction. The achievement of a scientist from a defeated and humiliated country had been recognized by the proudest of the victor nations. Furthermore, the discovery was not based on any collection of empirical researches but arose rather from a creative imagination which by its power had guessed the secret of the universe as it actually is, and the correctness of the solution of the puzzle had been confirmed by the precise astronomical observations of the cool-headed Englishmen.

The situation contained still another peculiarity in that the

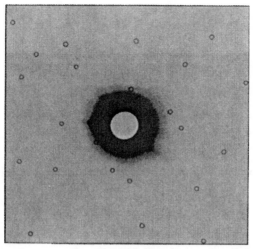

FIGURE 3

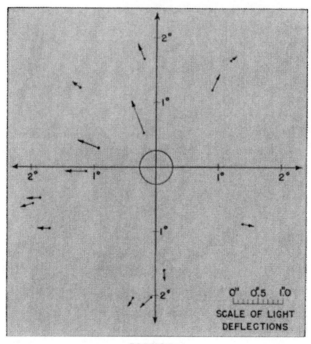

SCALE OF LIGHT
DEFLECTIONS

FIGURE 4

The observations of the British eclipse expedition of 1919 were repeated with
refined methods by an American expedition in 1922 at Wallal (Western Aus-
tralia) organized by the Lick Observatory, University of California. Fig. 3
shows the original photo of the eclipsed sun, its corona and the brightest stars in
the sun's vicinity. The images of the stars are encircled. Fig. 4 shows the observed deflection of the stars in the gravitational field of the sun. The arrows
represent the deflection of the images of the fifteen best measured stars in size
and direction. (Lick Observatory Bulletin 397)

Einstein and Charles Proteus Steinmetz

discoverer was a descendant of the Jewish people who had often been insulted and slighted by the defeated nation. The members of the Jewish community had often been compelled to hear and to read that while their race possessed a certain craftiness in business pursuits, in science it could only repeat and illuminate the work of others, and that truly creative talents were denied them. That this unique, ancient people had again produced a leader in the intellectual world not only seemed exciting for the Jews themselves, but also was a kind of consolation and stimulus for all the vanquished and humiliated people of the world.

A Russian observer gave this description of the remarkable psychological situation in defeated Germany at that time:

"With the growing social misery there appeared among the intellectuals pessimistic currents of thought, ideas about the decline of Western culture, and, with the violence of a hurricane, religious movements. The extent of these movements must seem remarkable even to one who is acquainted with German intellectual life. The number of independent religious groups in Germany grew endless. World War invalids, merchants, officers, students, artists, all were seized by the desire to create a metaphysical basis for their view of the world."

This flight from tragic reality into a dream world also increased the enthusiasm for Einstein's theory, which occupied a special place because it appeared to the public that here a portion of the reality of the universe had been discovered by dreams.

In Soviet Russia people at this time were in the process of constructing a new social order on principles that were consciously opposed to the pessimistic ideas of the "declining" West. They renounced all idealistic dreams or at least believed that they were doing so. They wanted to dissociate themselves as completely as possible from the attitudes prevalent both in the defeated and in the victorious countries. Everywhere they looked for signs of "decline." It was thought that such symptoms likewise evidenced themselves in the development of physical science. As early as 1922 A. Maximov, a leading exponent of Soviet Russian political philosophy who occupied himself especially with the physical sciences, wrote in the official philosophical journal of Soviet Russia in conjunction with the above description of German life:

"This idealistic atmosphere has surrounded and still surrounds the relativity theory. It is only natural, therefore, that the announcement of 'general relativity' by Einstein was received with delight by the bourgeois intelligentsia. The impossibility within the limits of bourgeois society for the intellectuals to withdraw from these influences led to the circumstance that the relativity principle served exclusively religious and metaphysical tendencies."

Here we note something of the feeling against Einstein that was to develop in some groups of the Soviet Union.

In this connection, however, it should not be forgotten that at the same time in Germany opinions were expressed characterizing Einstein's theory as "Bolshevism in physics," similar to those of the aforementioned American scientist. The rejection of Einstein's theories by some prominent Soviet spokesmen did nothing to change these opinions. And since the Bolsheviks and the Jews were commonly regarded as somehow related, we are not surprised to find that the relativity theory soon began to be regarded as "Jewish" and capable of harming the German people. This hostile attitude toward Einstein emanated in Germany from those circles which ascribed the loss of the war to the "stab in the back" and not to the failure of the ruling classes.

For Einstein himself this intrusion of politics and nationalism into the judgment of his theories was completely astonishing — indeed, hardly comprehensible. For a long time he had hardly paid any attention to these things and had not even noticed many such attacks. But gradually complete absorption in the regularities of the universe began to be difficult for him. More and more the anarchy of the human world pushed into the foreground. With brutal force it slowly but surely laid claim to a greater or lesser part of his intellectual energy.

VII

EINSTEIN AS A PUBLIC FIGURE

---

1. *Einstein's Political Attitude*

With the intense public interest aroused by the confirmation of his theory, Einstein ceased to be a man in whom only scientists were interested. Like a famous statesman, a victorious general, or a popular actor, he became a public figure. Einstein realized that the great fame that he had acquired placed a great responsibility upon him. He considered that it would be egoistic and conceited if he simply accepted the fact of his recognition and continued to work on his researches. He saw that the world was full of suffering, and he thought he knew some causes. He also saw that there were many people who pointed out these causes, but were not heeded because they were not prominent figures. Einstein realized that he himself was now a person whom the world listened, and consequently he felt it his duty to call attention to those sore spots and so help eradicate them. He did not think of working out a definite program, however, he did not feel within himself the calling to become a political, social, or religious reformer. He knew no more about such things than any other educated person. The advantage he possessed was that he could command public attention, and he was a man who was not afraid, if necessary, to stake his great reputation.

It was always clear to him that anyone venturing to express his opinion about political or social questions must emerge from the cloistered halls of science into the turmoil of the market place, and he must expect to be opposed with all the weapons common to the market place. Einstein accepted this situation as self-evident and included in the bargain. He also realized that many of his political opponents would also become his scientific opponents.

In the years immediately following the World War it was only natural that the main problem of all political reformers was the prevention of another such catastrophe. The obvious means to this goal were the cultivation of international conciliation, struggle against economic need, for disarmament, and the emphatic rejection of all attempts to cultivate the militaristic spirit. The surest and indeed an infallible method of obtaining the desired end seemed to be the refusal of military service by the individual, the organization of "conscientious objectors" on a large scale. All these ideas appeared as obvious to Einstein as they did to so many others. Only he had more courage and more opportunity than others to advocate them. Einstein did not have the self-complacency with which scholars, especially in Germany, liked to retire into the ivory tower of science. But the means toward the goal appeared to him at that time, as to many thousands, much simpler and more certain than was later found to be the case.

Einstein's political position, like that of all the intellectuals in the world, changed during the twenty years of armistice between the two World Wars, but he was never a member of any political party. Parties made use of his authority where they could do so, but he was never active in any group. This was due fundamentally to the fact that Einstein was never really interested in politics.

Only to very superficial judges does Einstein appear to be a genius so buried in his researches that he finds all his happiness in them without being influenced by the outside world. There are many more unresolved contradictions in Einstein's character than one would believe at first glance, and these, as I have mentioned already, are due to the contrast between his intense social consciousness on the one hand and the aversion to entering into too intimate relations with his fellow men on the other.

This trait manifests itself above all in his attitude toward political groups, with which he has co-operated at times because he sympathized with some of their aims. There were always moments when it was extremely vexatious for him to be forced into actions and expressions of which he did not approve, and the moment always recurred when he developed antipathies to the representatives of the groups with which he sympathized. Moreover, he did not like to claim any special role for himself and so he sometimes participated in things that were actually not very much to his liking. When something of this sort happened, naturally he did not become any fonder of the people who had caused him to do so. As a result he impressed many people as a vacillatory supporter. He always stood first for what seemed valuable to him, but he was not ready to let himself be influenced too much by party stereotypes and slogans. This was his attitude in his co-operation with the Zionists, pacifists, and socialists.

Einstein realized very well that everything has several aspects and that by supporting a good cause one must often help one that is less worthy. Many people who are essentially hypocrites seize upon such situations and refuse to participate in any good cause because of "moral scruples." Such behavior was not Einstein's way of acting. If the basic cause was good he was occasionally ready to take into the bargain a less worthy, secondary tendency. He was much too realistic and critical a thinker to believe that any movement conducted by human beings to attain human aims could be perfect.

He helped the Zionist movement, for instance, because he believed that it was of value in creating a feeling of self-respect among the Jews as a group and in providing a refuge for homeless Jews. He was well aware, however, that at the same time he was helping occasionally the development of nationalism and religious orthodoxy, both of which he disliked. He saw that at present no other instrument than a kind of nationalism was available to produce a feeling of self-respect in the rank and file of the Jewish community.

There were times, however, when the prospect of having his remarks interpreted falsely appeared so unpleasant to Einstein that he did not permit himself to be placed in a position where such a situation could develop. Einstein received repeated invitations to visit and lecture in Soviet Russia, especially during the early years of the development of her science, but he declined. Einstein realized that any friendly remark he might make to the country would be interpreted by the outside world as a sign that he was a Communist, and any critical remark would be taken by the Communists as a part of a capitalistic crusade against Russia.

2. *Anti-Semitism in Postwar Germany*

After the war, when Germany's defeat led to a collapse of the rule of the generals and the junkers, who had generally been regarded as the source of all prejudice, many people thought that the period of discrimination against the Jews was now past. But actually the loss of power aroused a deep-seated feeling of anger in these classes. A human being is inconsolable over a catastrophe only so long as he believes its cause was due to his own inferiority. Consequently he tries to put the blame on someone else. Thus the supporters of the overthrown rulers spread the idea that the defeat had been caused not by military weakness, but by an internal revolt led by the Jews. The spread of this view caused a feeling of extreme hatred against the Jews in Germany. Such sentiments were very widespread even among the educated class, and they were all the more dangerous for the Jews because they were completely irrational. The Jews could not refute them by any arguments or escape the enmity by any change in their conduct.

Many of the Jews in Germany, however, did not understand this situation, and they made efforts to divert attention from themselves through various kinds of mimicry. In the mildest form, they tried to shift the blame for the defeat by putting it on the lack of patriotism among the Socialists. Many went even further, emphasized a division among the Jews, and accused the "bad" group. The Jews who had been long resident in Germany ascribed all the inferior characteristics to the Jews who had immigrated from eastern Europe. Among them were included, depending on preference and momentary need, Jews from Poland, Russia, Rumania, Hungary, and sometimes even of Austria. When Hitler, who, as is well known, came from Austria, began his persecutions of the Jews, a Jewish professor at a German university said: "One cannot blame Hitler for his views about the Jews. He comes from Austria and he is right as far as the Jews there are concerned. If he had known the German Jews well, he would never have acquired such a poor opinion of us." Such statements characterize in drastic fashion the feeling of some German Jews. This feeling was so bitterly resented by the eastern Jews that when Hitler began to persecute the German Jews, the reaction was not a united front of all Jews but often one regional group tried to put the blame for Hitler on other Jewish groups.

This lack of self-respect in the behavior of some German Jews made a mortifying impression on Einstein. Until then he had taken little interest in the condition of the Jews and had hardly realized the grave problems in their situation, but now he developed a deep sympathy for their position. Although Einstein had a certain aversion to Jewish orthodoxy, he looked upon the Jewish community as a group that was the bearer of a very valuable tradition and that regarded intellectual values very highly. Hence, he saw with bitter feeling the Jewish community not only attacked by external enemies but also disintegrating inwardly. Einstein saw the Jews move deeper and deeper into a distorted psychological situation that could only produce a perverted mentality.

This profound sympathy aroused in him an ever increasing feeling of responsibility. As his fame grew, he gave the entire Jewish community the certainty that it was capable of producing a man with the creative intellectual power to formulate a theory of the universe recognized by the whole world as one of the greatest achievements of our time. Here was a refutation of the widespread opinion that truly creative intellectual powers are restricted to the Nordic-Aryan race.

---

3. *The Zionist Movement*

During the World War, when the British government declared its willingness to support the development of a national home for the Jews in Palestine, the Zionist movement experienced a powerful revival in all countries. Its goal was to establish a Jewish state in the ancient historical homeland of the Jews in order to give the Jews of the entire world a national and cultural center. In the British promise they saw the first step toward this goal. It was hoped that the co-operation of all the Jews in the world would enable them to throw off the humiliating feeling that they alone among all people had no national home and were everywhere tolerated only as guests.

From the beginning Einstein had various doubts about the Zionist aims. He was not sympathetic to the strong nationalistic emphasis, and he saw no advantage in substituting a Jewish for German nationalism. He also realized the difficulties inherent in the Palestine plan. He thought the country was too small to receive all the Jewish immigrants who might want to settle in a national home, and he foresaw the clash between Jewish and Arab nationalism. Zionists often have tried to minimize the magnitude of these problems, but Einstein considered this due to wishful thinking.

But in spite of all these doubts and scruples, Einstein saw many reasons in favor of Zionism. He saw in it the only active movement among the Jews that appeared capable of arousing in them the sense of dignity, the absence of which he deplored so much. He did not much care to have this educational process put into effect by an emphasis on nationalism, yet he felt that the Jewish psyche, and in particular that of the German Jews, was in such a pathological state that he recommended every educational means that tended to alleviate and remedy this situation.

He therefore decided in 1921 to appear publicly as a supporter of Zionism. He was well aware that this action would produce an astounding impression within German Jewry, since almost all the Jews in Germany who were active in public life as scholars and writers considered the Zionist movement as a mortal enemy of the development that they sought — the gradual complete assimilation of the Jews among their fellow citizens. When a man like Einstein, certainly the greatest of the Jewish scientists in Germany and a man of world reputation, stepped forth in this manner and thwarted their efforts, it was obvious that by many German Jews his action would be regarded as a "stab in the back." But Einstein was not the man to be afraid of any such reaction. He even felt that this antagonism was already the beginning of the educational process at which he was aiming. Also, since Einstein had taken upon himself to say so much that other people did not dare express, self-expression became easier and inhibitions were abated.

Thenceforth Einstein has been regarded by many people as a "black sheep" among the German scholars of Jewish origin. Attempts were made to explain his conduct on the basis of all sorts of causes, such as his misunderstanding of the German character, his wife, the propaganda of skillful journalists, or even his being, allegedly, a "Russian draft-dodger." They did not realize that Einstein was utilizing the credit he had obtained through his scientific achievements to educate the Jewish community.

Einstein's participation in the work of the Zionists, however, was due not only to the primary aim of this movement, but as well to a secondary plan that struck a responsive chord in his innermost being. This was the plan to establish a Jewish university in Jerusalem.

It had always been very painful for Einstein to see many Jewish youths who wished to acquire a higher education prevented from doing so on account of the discrimination against them. Most universities in eastern Europe were averse to admitting a large number of Jewish students. In central Europe, again, this attitude prevailed against the admission of Jewish students barred from the eastern universities. To Einstein it appeared as a special form of brutality — indeed, a paradoxical brutality — that just these people who had always had a special respect and love for intellectual pursuits should be thwarted in their ambitions. Although the Jewish students were often among the most interested and industrious, every admission of an Eastern Jew to a Germanic university was regarded as a special act of tolerance. Thus even the few fortunate ones who were admitted were not fully regarded by the others as fellow students and friends, and they never felt really at ease. The same prejudice was also felt by quite a few Jewish teachers. For this reason Einstein felt that it was necessary to found a university that would belong to the Jews and where students and professors would be free of the tension that arises through constant contact with an unfriendly environment.

It was through this plan for a university that Einstein came into contact with Chaim Weizmann, the recognized leader of the Zionist movement. Like Einstein, Weizmann was a scientist, but he was more interested in the application of science to technical problems. He was a professor of chemistry at the University of Manchester in England, and his work in the war research had been of great service to the British government during the World War. As a result he had become associated with influential English circles and had thus been able to propagate the Zionist plan. Einstein certainly intended to collaborate with the party led by Weizmann for a definite purpose, and the plan for the establishment of a university in Jerusalem rendered this collaboration easier. Weizmann himself characterized the aims of the university in a far-sighted way that Einstein must also have found sympathetic. He said: "The Hebrew University should further self-expression and shall play a part as interpreter between the Eastern and Western world."

4. *Einstein as a Pacifist*

From his childhood Einstein had been terribly depressed at the sight of people being trained to become automatons, whether they were soldiers marching through the streets or students learning Latin at the gymnasium. Aversion to mechanical drill was combined in him with an extreme abhorrence of all violence, and he saw in war the culmination of all that was hateful — mechanized brutality.

Einstein placed this aversion above and apart from any political conviction. On one occasion, speaking to a group of Americans who visited him in Berlin in 1920, he said:

"My pacifism is an instinctive feeling, a feeling that possesses me because the murder of men is disgusting. My attitude is not derived from any intellectual theory but is based on my deepest antipathy to every kind of cruelty and hatred. I might go on to rationalize this reaction, but that would really be *a posteriori* thinking."

Because Einstein's attitude to war was based on general human grounds rather than on political ones, it was very difficult for him to work together with institutions that also considered themselves to be working for world peace. In 1922 Einstein was appointed to the "Commission pour la Coopération Intellectuelle" of the League of Nations. The purpose of this body was to make intellectuals acquainted with the aims of the League and to induce them to use their knowledge and talents for the achievement of these aims. The commission never got beyond certain vague beginnings. At first, however, Einstein believed that he ought not to refuse to co-operate, and in his letter of acceptance he wrote as follows: "Even though I must admit that I am not at all clear as to the character of the work to be done by the commission, I consider it my duty to obey its summons since nobody in these times should refuse assistance to efforts toward the realization of international co-operation."

But after one year Einstein recognized that the League did not prevent the use of force by great powers and sought only for means to induce weak nations to submit without resistance to the demands of the strong ones.

Consequently he resigned from the commission, giving the following reason for his action: "I have become convinced that the League possesses neither the strength nor the good will necessary to accomplish its task. As a convinced pacifist it does not seem well to me to have any relation whatever with the League."

In a letter to a pacifistic magazine he presented an even sharper formulation of this step:

"I did so because the activities of the League of Nations had convinced me that there appeared to be no action, no matter how brutal, committed by the present power groups against which the League could take a stand. I withdrew because the League of Nations, as it functions at present, not only does not embody the ideal of an international organization, but actually discredits such an ideal."

The correctness of his judgment was shown already in the fall of that year (1923) when in the conflict between Greece and Italy the League endeavored only to make Greece, the weaker party, yield. It did not wish to hurt the feelings of Italy, which was then celebrating the honeymoon of Fascism.

Soon, however, Einstein realized that the matter had another aspect. He noticed that his resignation from the commission was greeted with glee by the German nationalist groups. Thereupon, as on so many other occasions, he reflected that even though one sees many mistakes in a movement, yet it is not right to refuse to support it if its essential principle is a good one. In 1924 he therefore rejoined the commission. On the occasion of the tenth anniversary of the League (1930) he expressed the essence of his opinion as follows: "I am rarely enthusiastic about what the League has done or has not done, but I am always thankful that it exists." He always emphasized, however, that without the collaboration of the United States the League would never become a factor for international justice.

Einstein always thought that scientists have a special part to play in advancing the cause of international understanding. The nature of their work is not restricted by national boundaries as is the case, for instance, with history and economics, and their judgments of merit tend to be objective. It is therefore particularly easy for scientists of different countries to find a common ground. As he once put it:

"The representatives of the natural sciences are inclined, by the universal character of the subject dealt with and by the necessity of internationally organized co-operation, toward an international mentality predisposing them to favor pacifistic objectives. . . . The tradition of science as a force in cultural training would open a much more comprehensive view before the mind and would be a powerful influence — because its outlook is world-wide — in drawing men a little way from senseless nationalism. You cannot drive out nationalism unless you put something in its place. And science gives this wide something which men could hang on to."

Here Einstein also saw a task for the Jewish people. For centuries the Jews had formed such a small minority among their neighbors that they had been unable to defend themselves by physical force. They had shown how in the face of physical violence it is possible to survive by intellectual means. In an address at a Jewish meeting in Berlin in 1929 Einstein said:

"Jewry has proved that the intellect is the best weapon in history. Oppressed by violence, Jewry has mocked her enemies by rejecting war and at the same time has taught peace. . . . It is the duty of us Jews to put at the disposal of the world our several-thousand-years-old sorrowful experience and, true to the ethical traditions of our fore-fathers, become soldiers in the fight for peace, united with the noblest elements in all cultural and religious circles."

Einstein's attitude to pacifism must always be kept in mind if one wants to understand his political position. As the prob-lem of social reorganization became more and more complicated and it was no longer certain which groups represented progress toward this goal, Einstein refused definitely to link the fight against war with the cause of socialism. The American Socialist leader Norman Thomas once asked him whether he did not consider the realization of a socialist society a necessary prereq-uisite to guarantee general peace. Einstein replied:

"It is easier to win over people to pacifism than to socialism. Social and economic problems have become much more difficult today, and it is necessary that men and women reach the point where they believe in pacific solutions. Then you can expect them to approach economic and political problems in a spirit of co-operation. I would say that we should not work for socialism first, but for pacifism."

Just as Einstein was aware that social problems cannot be solved by a simple declaration of faith in socialism, but that very complicated and often antithetical interests must be reconciled, so he had likewise long been cognizant of the paradox in the ideal of democracy. The people should rule, yet freedom can never be realized by means of a formula, but only if the system is headed by men worthy of the confidence placed in them. Democracy necessarily leads to the formation of parties; but mechanical party rule often leads in turn to the suppression of oppositional groups. Thus he wrote in 1930:

"My political ideal is democracy. . . . However, well do I know that in order to attain any definite goal it is imperative that *one* person should do the thinking and commanding and carry most of the respon-sibility. But those who are led should not be driven, and they should be allowed to choose their leader. It seems to me that the distinctions separating the social classes are false; in the last analysis they rest on force. I am convinced that degeneracy follows every autocratic system of violence, for violence inevitably attracts moral inferiors. Time has proved that illustrious tyrants are succeeded by scoundrels."

Einstein never considered the essence of democracy to be the observance of certain formal rules; it lay chiefly, rather, in the absence of any spirit of violence directed against certain sections of the nation. Even before Germany became a dictatorship, he had already recognized the shady sides of this system, as well as those of the still prevailing formal democracy. He once said:

"For this reason I have always been passionately opposed to such regimes as exist in Russia and Italy today. The thing that has discredited the European forms of democracy is not the basic theory of democracy itself, which some say is at fault, but the instability of our political leadership, as well as the impersonal character of party alignments."

At that time Einstein already regarded the American system of government as a form of democracy superior to the German or even the French republic. It was based not so much on parliamentary deliberations and votes as on the leadership of an elected president. "I believe," Einstein told an American journal in 1930, "that you in the United States have hit upon the right idea. You choose a president for a reasonable length of time and give him enough power to acquit himself properly of his responsibilities."

Similarly, during the discussions over Roosevelt's third term Einstein was unable to agree with the view that the number of terms to which a president is elected is important for democracy, because he felt that the spirit in which the president exercised the powers of his office was much more significant.

But while the question of democracy or socialism always seemed complicated to him and incapable of solution by a formula, yet at that time the problem of his attitude to military service and war still seemed simple because his aversion was not based on any political convictions.

It is even possible to find statements by Einstein that sound "undemocratic" and almost like an espousal of the doctrine of élite, as, for instance: "What is truly valuable in our bustle of life is not the nation, I should say, but the creative and impressionable individuality, the personality — he who produces the noble and sublime while the common herd remains dull in thought and insensible in feeling." And he hated all military institutions because they cultivated and developed this very herd spirit:

"This subject brings me to the vilest offspring of the herd mind — the odious militia. The man who enjoys marching in line and file to the strains of music falls below my contempt: he received his great brain by mistake, the spinal cord would have been amply sufficient. This heroism at command, this senseless violence, this accurate bombast of patriotism — how intensely I despise them!"

Einstein was not opposed to a dictatorship because it recognized the existence of an élite, but because it tried to develop a herd mind among the majority of the people.

This goal — the avoidance of war and military service — seemed so desirable that in this case he believed the most primitive and most radical means to be the most effective — that is, the refusal of the individual to perform military service, as practiced by certain religious groups such as the Quakers or Jehovah's Witnesses. In 1929, when he was asked what he would do in case of a new war, he replied in a magazine: "I would unconditionally refuse to do war service, direct or indirect, and would try to persuade my friends to take the same stand, regardless of how the cause of the war should be judged." In 1931 he placed his reputation and his personal co-operation at the disposal of the War Resisters International and issued an appeal in which he said:

"I appeal to all men and women, whether they be eminent or humble, to declare that they will refuse to give any further assistance to war or the preparation of war. I ask them to tell their governments this in writing and to register this decision by informing me that they have done so. . . . I have authorized the establishment of the 'Einstein War Resisters International Fund.'"

When I visited the House of Friends in London, the headquarters of the Quakers, I saw the pictures of three men in the secretary's office: Gandhi, Albert Schweitzer, and Einstein. I was rather surprised at this combination and asked the secretary what it was that these three persons had in common. Amazed at my ignorance, he informed me: "All three are pacifists."

---

5. *Campaigns against Einstein*

The German intellectuals who had blindly followed the ruling military class into the first World War were bewildered when their trust was broken by the loss of the war. The professors in the years immediately following the armistice felt like sheep without a shepherd. When Einstein ventured forth into this confused atmosphere by entering into public affairs in support of Zionism and pacifism, strong opposition began to be organized against him.

For the ardent nationalists, the Jews and the pacifists were the scapegoats to be blamed for the defeat in the war by a "stab in the back," and any supporter of their movements was the object of their violent anger. Even those who agreed with Einstein's ideas were shocked by his blunt way of speaking in face of the opposing sentiment, and he began to be looked upon as a kind of *enfant terrible*. Einstein was not familiar with political machinations and had no interest in them, and so his statements were considered either childish or cynical. With the success of his theory acclaimed by the English solar expedition and the rise of his fame, his enemies set out to depreciate this success as far as possible.

There suddenly appeared an organization whose only purpose was to fight Einstein and his theories. The leader was a certain Paul Weyland, whose past, education, and occupation were unknown. The organization had at its disposal large sums of money of unknown origin. It offered relatively large fees to people who would write against Einstein or oppose him at meetings. It organized meetings by means of large posters, such as were used to announce the greatest virtuosi.

The people who spoke for and represented this movement may be divided into three groups. The first group comprised the political agents of the "revolution from the Right." They knew absolutely nothing about Einstein and his theories except that he was a Jew, a "pacifist," that he was highly regarded in England, and that he also seemed to be trying to gain a hold on the German public. These people spoke loudest and with the greatest assurance. As professional propagandists usually do, they accused Einstein and his supporters of making too much propaganda. They did not enter into any objective discussions, and they intimated in a more or less veiled fashion that the spread of Einstein's theory was due to the same conspirators who were to blame for the German defeat. Since it is characteristic of the mode of thought of this group, I wish to quote from an article that appeared in the magazine *Der Türmer,* a literary monthly that was highly regarded in German nationalistic circles. Under the title "Bolshevistic Physics" Einstein's theory is directly related to the political situation. In the opinion of many, the German defeat was due to the circumstance that President Woodrow Wilson had promised the Germans a just peace, and had thus led them to conclude an armistice, as they were not compelled to do by the military situation. The article continues:

"Hardly had it become clear to the horrified German people that they had been frightfully duped by the lofty politics of Professor Wilson and swindled with the aid of the professorial nimbus, when a new professorial achievement was again being commended to the simple Germans with the greatest enthusiasm and ecstasy as the pin-nacle of scientific research. And unfortunately even highly educated people fell for this — all the more so since Professor Einstein, the alleged new Copernicus, numbers university teachers among his ad-mirers. Yet, without mincing words, we are dealing here with an infamous scientific scandal that fits very appropriately into the picture presented by this most tragic of all political periods. In the last analysis one cannot blame workers for being taken in by Marx, when German professors allow themselves to be misled by Einstein."

A second group in this movement directed against Einstein was composed of several physicists who had acquired a reputa-tion in professional circles as a result of precise experiments, and who now wondered that someone could become world-famous because of the constructions of his creative imagination. They lacked the comprehensive vision to realize the necessity for such far-reaching generalizations as those of Einstein; on the whole they were inclined to see only that honest hard-work-ing physicists were being slighted in favor of a frivolous in-ventor of fantastic hypotheses. Here there was already some-thing of the idea that the ability to observe nature faithfully was a characteristic of the "Nordic" race, which Einstein conse-quently lacked.

The third group was composed of philosophers who advo-cated certain philosophical systems that were inconsistent with the theory of relativity. Or, to put it more precisely, they did not understand the exact physical meaning of the relativity theory, and so they attributed to it a metaphysical interpretation that it actually did not contain. Then they denounced this phi-losophy that they themselves had invented. Here, too, there is already something of the conception that the Nordic-Aryan phi-losopher probes the true profundity of nature itself while other races are satisfied with a discussion of how nature may be de-scribed from different points of view.

But since physicists as well as philosophers are often very naïve or, to put it more plainly, are very thoughtless in matters of individual and political psychology, the latter two groups were frequently not even aware that they were acting in the service of a specific political propaganda.

When Paul Weyland organized his first meeting in the Berlin Philharmonie, he even made great efforts to secure speakers who were of Jewish origin so as to create a kind of smoke screen. At this first meeting Weyland, whose speech was more political than scientific, was followed by E. Gehrcke, a competent experimental physicist of Berlin, who criticized the theory from the point of view of a man who, while making no mistakes in his experiments, simply lacks the acute understanding and flight of imagination to pass from individual facts to a synthesis. Such people are usually ready to accept old hypotheses because through habit they have forgotten that they are not facts, but they like to stamp new theories as "absurd" and "opposed to the spirit of empirical science." An invitation had also been extended to a representative of philosophy who was to prove that Einstein's theory was not "truth," but only a "fiction." He was of Jewish descent and was intended to be the climax of the meeting. Despite his political innocence and urgent telegrams, he declined at the last moment because some friends had explained the purpose of the meeting to him. As a result the first attack took place without the blessing of philosophy.

Einstein attended this meeting as a spectator and even applauded the attacks in a friendly spirit. He always liked to regard events in the world around him as if he were a spectator in a theater. The meetings of this group were just as amusing as the sessions of the university faculty in Prague, or of the Prussian Academy of Science.

Other meetings of this group took place, and in this year the "Einstein case" became a constant subject of discussion in the press. Einstein was besieged with requests to express publicly his opinion of these attacks. But it was repugnant to him to act as if he thought that he was dealing with scientific discussions. He had no desire to discuss publicly questions that for most people were incomprehensible and that played no part whatsoever in these meetings. Finally, in order to terminate the entire affair, he wrote in a Berlin newspaper that it would be senseless to answer in a scientific manner arguments that were not meant scientifically. The public would not be able to judge who was right. Therefore he said simply: "If I were a German nationalist with or without a swastika, instead of being a Jew with liberal, international opinions, then . . ." This was more understandable to everyone than scientific arguments would have been.

Now Einstein's opponents were more furious than ever and asserted that Einstein was turning a scientific discussion into a political one. Actually, he had once again been the *enfant terrible* and had simply called the thing by its right name. Even many of his friends would have preferred it had he acted as if he did not understand the motives of his opponents.

At this time he began to feel uneasy in Berlin and there was a great deal of talk that he would leave Germany. He was also offered a professorship at the Dutch University of Leiden. When he was asked whether he actually wanted to leave Berlin, he said: "Would such a decision be so amazing? My situation is like that of a man who is lying in a beautiful bed, where he is being tortured by bedbugs. Nevertheless, let us wait and see things develop."

The movement against Einstein acquired a certain respectability as soon as a physicist who was generally regarded as an outstanding member of his profession put himself at its head. I have already spoken of Philipp Lenard on several occasions. In 1905 Einstein had based his new conception of light on the observations of Lenard. For these and other experiments carried out with the greatest ingenuity, Lenard had received the Nobel prize. He was less skilled, however, in deriving general laws from his observations. When he tried to do so, he involved himself in such complicated hypotheses that they could not contribute to any clarification. He therefore achieved no recognition as a theoretician.

He was one of the physicists who during the World War had become extreme nationalists and particularly embittered enemies of England. He regarded the defeat, which came quite unexpectedly for him as well as for the others who held the same political views, as the work of international powers: namely, the Socialists and pacifists. He was one of those who began to accuse the Jews of being the actual wire-pullers in the background. Lenard soon joined the Hitler groups and was a very old member of the National Socialist Party.

He was astonished that Einstein had such great success after the war. In the first place, this man was not an experimenter; secondly, he was the inventor of an "absurd" theory that contradicted sound common sense as embodied in mechanistic physics; and thirdly, in addition to all this he was a Jew and a pacifist. For Lenard this was more than he could stand, and he put his reputation and prestige as a physicist at the service of Einstein's opponents. In him were united the motives of all three

162

groups that opposed Einstein: the agents of the revolution from the Right, the pure "empiricists," and the advocates of a certain philosophy.

Lenard's nationalistic fanaticism was revealed by many incidents. On one occasion the well-known Russian physicist A. F. Joffe was traveling through Germany after the war in order to resume contact with German colleagues. He went to Heidelberg and wanted to visit Lenard to discuss scientific subjects with him. He requested the porter at the institute to announce him to Lenard. The porter returned and said to Joffe: "Herr Geheimrat Lenard wishes me to say that he has something more important to do than to converse with the enemies of his fatherland."

As is well known throughout the entire world the unit of intensity of an electrical current is called an ampere, after the French physicist and mathematician André M. Ampère. Lenard, however, ordered that in his laboratory the electrical unit must change its French name and assume the name of a German physicist, Weber. This change was made on all the instruments in the Heidelberg laboratory.

Every year in September there was a meeting of German-speaking scientists and teachers of sciences. Usually several thousand persons came together. In 1920 this meeting was to take place at the well-known spa Bad Nauheim. Several papers dealing with the relativity theory were also on the program. Lenard decided to take this opportunity to attack Einstein's theories before the assembled scientists and to demonstrate their absurdity.

This became generally known and the session was awaited as if it were a sensational and decisive meeting of a parliament. Max Planck presided. This great scientist and distinguished man detested any kind of sensation. He endeavored to arrange the session in such a way so as to keep the debate on the level on which scientists usually discuss matters and to prevent non-scientific points of view from being brought into it. He arranged it so that the greatest part of the available time was filled with papers that were purely mathematical and technical. Not much time remained for Lenard's attack and the debate that would ensue. The entire arrangement was made to prevent any dramatic effects.

Questions of principle were not touched upon in the long reports, which were full of mathematical formulæ. Then Lenard took the floor for a short talk in which he attacked Einstein's theory, but without introducing any emotional coloring. His argument was neither that the theory was inconsistent with experimental results, nor that it contained logical contradictions, but actually only that it was incompatible with the manner in which ordinary "common sense" conceived things. Fundamentally it was only a criticism of a language that was not that of mechanistic physics.

Einstein replied very briefly and then two others spoke still more briefly for and against Einstein. With this the session came to an end. Planck was able to heave a sigh of relief that the meeting had passed without any major conflict. The armed policemen who had watched the building were withdrawn. Planck was in such good humor that he ended the session with one of the trivial jokes that have been current among non-physicists: "Since the relativity theory unfortunately has not yet made it possible to extend the absolute time interval that is available for the meeting, our session must be adjourned."

To a certain degree the lack of understanding of the philosophical significance of Einstein's theory among most professional physicists stood in the way of a real debate in which it would have been possible to explain the true content of the theory to its well-meaning opponents. As a result, however, the impression was created that while the Einstein theory might have a meaning for mathematicians, yet for a more philosophically thinking mind it contained various absurdities.

Thus Lenard himself received the impression that not enough attention was paid to his arguments, and the mass of physicists and mathematicians had no opportunity to take part in a truly fundamental discussion on a grand scale. For the moment the physicists probably felt relieved that nothing worse had happened; nevertheless, the opportunity had been permitted to pass without organizing a real explanation for the great mass of scientists and educated persons.

The opposition of Lenard and his supporters to Einstein's theory was checked by one fact. Even though the foundations of the theory could be characterized as "absurd" and "muddled," yet it was undeniable that inferences could be drawn from this "absurd" theory that every scientist had to admit were usable and important. Even the most vigorous opponent, if he was a physicist or chemist, had to reckon with the formula that represented the relation between mass and energy. If the energy E is given off, this is equivalent to a loss of mass E/c^2, where c is the velocity of light (see Chapter III). Even the most zealous adherent of the "revolution from the Right" had to use this formula $E = mc^2$ if he wanted to penetrate the nucleus of the atom. Consequently Lenard and his group endeavored to separate this law from its connection with Einstein's theory and to prove that it had already been known before Einstein, having been advanced by a physicist of whose racial origin and sentiments they approved.

In the writings of those who want to avoid the name of Einstein at any price, the law of the transformation of mass into energy is often to be found as the "principle of Hasenöhrl." It is interesting, perhaps, for an understanding of the entire milieu in which Einstein worked to describe how this deliberate removal of Einstein's name occurred.

It had long been known that light falling on a surface exerts a pressure on it as if particles were being hurled against the surface. In 1904 the Austrian physicist Hasenöhrl had concluded from this knowledge: if light radiation is enclosed in a vessel, it will exert a pressure on the walls. Even if the vessel itself does not have any material mass, yet because of the pressure of the enclosed radiation it would under the impact of a force behave like a body with material mass. And this "apparent" mass is proportional to the enclosed energy. When the vessel radiates energy E the "apparent" mass m will decrease, according to $E = mc^2$.

This principle is obviously a special case of Einstein's law. If radiation is already contained in a body, its mass will decrease when the radiation is given off. Einstein's law, however, is much more general. He says that the mass of a body, no matter what its nature, decreases if the body loses energy in any manner whatsoever.

Lenard and his group, however, were seeking a substitute for the name Einstein. There were several external factors that favored the choice of Hasenöhrl's name. During the World War he had fought in the Austrian army — that is, on Germany's side — and was killed in battle at the age of forty. He thus appeared to be an ideal figure in the view of Einstein's enemies, a hero and model for German youth who was the very antithesis of the abstract speculator and international pacifist Einstein. Actually, Hasenöhrl was a honest and competent scientist and a sincere admirer of Einstein.

The legend originated with Lenard's book *Great Men of Science*. The author presented a series of biographies of great men, such as Galileo, Kepler, Newton, Faraday, and others, and he concluded with one of Hasenöhrl. In order to link him with the preceding heroes Lenard said of him: "He loved music and his violin as Galileo his lute: he was very fond of his family and as extremely modest as Kepler." Further on he says about Hasenöhrl's conclusions: "The applications of this idea have already progressed very far today, although almost entirely in the names of other people." "Other people" apparently means Einstein.

VIII

TRAVELS THROUGH EUROPE, AMERICA,

AND ASIA

---

1. *Holland*

The vicious attacks on Einstein resulted in arousing interest in his theories among all classes of people in every country. Theories that were of no great significance to the masses and at that almost incomprehensible to them became the center of political controversies. At a time when political ideals had been shattered by the war and new philosophies and political systems were being sought, the public was puzzled and mysteriously attracted by the connection between Einstein's scientific work and politics. The public interest was further increased by the appearance of articles by philosophers published in daily newspapers stating that Einstein's theories might perhaps be of some importance in physics, but were certainly untrue philosophically.

The public wondered what sort of man was this Einstein, and they wanted to see and hear this famous scientist in person. From every country Einstein began to receive invitations to come and give lectures. He was amazed, but happy to comply with the people's wishes. He enjoyed leaving the narrow circle of his professional colleagues and coming into contact with new people. It was also refreshing for him to leave Berlin and Germany, to go away from this tormented and harrowing atmosphere, and to see new countries.

These journeys and public appearances, however, added another cause for attacks on Einstein. Even some German scientists became annoyed, and one of them, a hard-working observer in the laboratory, wrote a brochure entitled *The Mass Suggestion of the Relativity Theory*. In it he gave an interpretation of Einstein's travels from his own point of view. He wrote: "As soon as the erroneous character of the relativity theory became evident in scientific circles, Einstein turned more and more to the masses and exhibited himself and his theory as publicly as possible."

The first instance of this "unscientific" publicity was Einstein's lecture at the ancient and honorable University of Leiden in Holland. There he lectured before fourteen hundred students of this famous center of physical science on "Ether and the Relativity Theory." This lecture led to many misinterpretations. Einstein, who had previously suggested that the term "ether" be dropped so as to prevent the rise of any idea that one is dealing with a material medium, discussed another proposal: namely, the word "ether" be used for "curved space," or what amounts to the same thing, for the gravitational field present in space.

Einstein's new proposal irritated some physicists and made others happy. Quite a few were unable to differentiate between a proposal to use a word in a certain sense and an assertion of a physical fact. They said: "For a long time efforts were made to convince us of the sensational fact that the ether had been got rid of, and now Einstein himself reintroduces it; this man is not to be taken seriously, he contradicts himself constantly."

Einstein, however, was happy to be in the quiet, pleasant city of Leiden, among good friends and remote from the controversies of Berlin. He loved to carry on discussions with a physicist of this city, Paul Ehrenfest, a Viennese by birth who was married to a Russian physicist. Husband and wife were indefatigably ready to discuss with Einstein the most subtle questions regarding the logical relations of physical propositions.

Einstein was also appointed professor at Leiden, but he was required to lecture there only a few weeks throughout the year. It was a pleasant thought to look forward to this period of rest every year. And in Berlin there were constant speculations as to whether Einstein would move permanently to Holland. His opponents tried everything possible to make it unpleasant for him to remain in Berlin. Many Germans thought that they must be thankful to Einstein because by means of his great popularity abroad he was acting to increase Germany's prestige after the lost war. His enemies, however, began a campaign against him, asserting that he was making propaganda abroad only for his own reputation, not for Germany.

Hänisch, the Prussian Minister of Education and a member of the Social Democratic Party, wrote an anxious letter to Einstein entreating him not to let himself be disturbed by these attacks and to remain in Germany. The government of the German Republic was very well aware how valuable he was for German culture and for its prestige throughout the world. The German government was even sorry that the great new theory of a German scientist had been studied and confirmed by English astronomers so that a large part of the ensuing fame had been lost to the Germans. The Minister requested Einstein to make use of the assistance of German observers and promised him governmental assistance.

Einstein, who appreciated greatly the significance of Berlin as a center of science and research, also understood very well that it was now important for all progressively minded elements to do everything possible to increase the prestige of the German Republic. He wrote a letter to the Minister in which he said: "Berlin is the place to which I am bound by the closest human and scientific ties." He promised if possible to remain in Berlin and even applied for German citizenship, something that he had not wanted to accept from the Imperial government. He thus became a German citizen, a circumstance that later entailed only troubles for him.

---

2. Czechoslovakia

In Prague, which was now the capital of the new Czechoslovak Republic, a Urania society had been organized to arrange lectures for the German-speaking population, and in particular to acquaint them with the great personalities of the new republican Germany. The president of Urania, Dr. O. Frankel, also endeavored to induce Einstein to lecture in Prague. Einstein, who was fond of recalling the quiet times he had enjoyed when he was working in Prague, seized the opportunity to revisit his old university and friends. He was also interested in becoming acquainted with the new democratic state that had arisen under the leadership of President Masaryk on the ruins of the Habsburg monarchy. The psychological status of the German minority in Prague and in Czechoslovakia in general was approximately like that of the population of the defeated German Reich within Europe. Einstein's visit increased the self-esteem of the Germans in Czechoslovakia, who were later called "Sudeten Germans," and played a fateful role in the crisis that led to the second World War. When Einstein's visit was announced, one of the papers of this minority wrote: "The whole world will now see that a race that has produced a man like Einstein, the Sudeten German race, will never be suppressed." This was characteristic of the nationalistic thinking. On the one hand every effort was made to keep the race free of all foreign admixtures; on the other hand, when someone was needed, even one who had not spent two years among this race was counted as a member of the group.

Early in 1921 Einstein returned to Prague, where I was then teaching as his successor. I had not seen him for years. I remembered only the great physicist, the man with an artistic and often jesting outlook on the world, who at that time had already enjoyed a great reputation among scientists. But during the years that had passed since then, he had become an international celebrity, a man whom everyone recognized from his photographs in the newspapers, whose opinions on politics and art were sought by every reporter, and whose autograph was wanted by every collector; in short, a man whose life no longer belonged entirely to himself. As so often happens in such cases, he had ceased to be an individual person in many respects; he had now become a symbol or banner upon which the gaze of masses of people was directed.

I was therefore very curious to meet him again, and was somewhat worried about how I could make it possible for him to live a halfway quiet life in Prague and prevent him from being overburdened by his obligations as a famous man. When I met him at the station, he had changed very little and still looked like an itinerant violin virtuoso, with the combination of childlikeness and self-assurance that attracted people to him, but that sometimes also offended them. I had been married only a short time then, and during this period shortly after the war it was so difficult to find an apartment that I lived with my wife in my office at the Physics Laboratory. It was the same room, with the many large windows looking out on the garden of the mental hospital, that had formerly been Einstein's office. Since Einstein would have been exposed to curiosity-seekers in a hotel, I suggested that he spend the night in this room on a sofa. This was probably not good enough for such a famous man, but it suited his liking for simple living habits and situations that contravened social conventions. We told no one about this arrangement, and no journalist or anyone else knew where Einstein spent the night. My wife and I spent the night in another room. In the morning I came to Einstein and asked him how he had slept. He replied: "I felt as if I were in a church. It is a remarkable feeling to awake in such a peaceful room."

We went first to police headquarters, where, as was common during the postwar period, every stranger had to report. Then we visited the Physics Laboratory of the Czech University. The professors there were pleasantly surprised by seeing Einstein, whose picture hung on the wall, appear in person in their room. By this visit Einstein wanted to express his sympathy for the new Czechoslovak Republic and its democratic policy under Masaryk's leadership.

In Prague, as in all the cities that had belonged to the Austro-Hungarian monarchy, a large part of the social life took place in the cafés. There people read newspapers and magazines, met with friends and acquaintances, and discussed business problems and scientific, artistic, or political questions. New political parties, literary circles, or large business firms were founded in cafés. Often, however, people sat alone, studying books or doing their own writing. Many students prepared for their examinations there, because their rooms were too cold, too dark, or simply too dreary. Einstein wanted to visit such places and he said to me: "We ought to visit several cafés and look in to see what the various places frequented by different social classes look like." Thus we paid rapid visits to several cafés; in one we saw Czech nationalists, in another German nationalists; here were Jews, there Communists, actors, university professors, and so forth.

On the way home Einstein said to me: "Now we must buy something for lunch so that your wife won't have too much bother." At that time my wife and I cooked our meals on a gas burner such as is used for experiments in chemical or physical laboratories, a so-called Bunsen burner. This took place in the same large room in which we lived and where Einstein had also slept. We came home bringing some calf's liver that we had purchased. While my wife began to cook the liver on the gas burner, I sat with Einstein talking about all sorts of things. Suddenly Einstein looked apprehensively at the liver and jumped at my wife: "What are you doing there? Are you boiling the liver in water? You certainly know that the boiling-point of water is too low to be able to fry liver in it. You must use a substance with a higher boiling-point such as butter or fat." My wife had been a college student until then and knew little about cooking. But Einstein's advice saved the lunch; and we got a source of amusement for all our married life, because whenever "Einstein's theory" was mentioned, my wife remembered his theory about frying calf's liver.

That evening he lectured before the Urania association. It was Einstein's first popular lecture that I had heard. The hall was dangerously overcrowded since everyone wanted to see the world-famous man who had overthrown the laws of the universe and proved the "curvature" of space. The ordinary public did not really know whether it was all a colossal humbug or a scientific achievement. Nevertheless, it was ready to marvel at both. As we were going in to the lecture, a very influential man in public life who had himself done a great deal to organize the meeting pushed through the crowd and said to me: "Please tell me quickly in one word, is there any truth in this Einstein or is this all bunk?" Einstein spoke as simply and clearly as possible. But the public was much too excited to understand the meaning of the lecture. There was less desire to understand, than to experience an exciting event.

After the lecture the chairman of the Urania gathered together a number of guests to spend the evening with Einstein. Several speeches were made. When Einstein's turn came to answer, he said: "It will perhaps be pleasanter and more understandable if instead of making a speech I play a piece for you on the violin." It was easier for him to express his feelings in this way. He played a sonata by Mozart in his simple, precise, and therefore doubly moving manner. His playing indicated something of his intense feeling for the complexity of the universe and simultaneously of the intellectual joy over the possibility of expressing it in simple formulas.

Einstein remained in Prague another evening to participate in a discussion of his theories that was to take place in the Urania before a large audience. Einstein's main opponent was a philosopher of the Prague University, Oskar Kraus, an acute thinker in the philosophy of law, whose conception of scientific discussions, however, was more like that of a counsel at a trial. He made no attempt to explore the truth, but instead wanted only to refute his opponent by finding passages that were contradictory in the writings of Einstein's supporters. In this he was successful. Anyone who wants to present a complex subject popularly must introduce some simplifications. But every author introduces them at different places according to his own taste or his opinion of his reader's tastes. If every statement by a popularizer is then taken literally, contradictions must necessarily arise. But this has nothing to do with the correctness of Einstein's theory.

Professor Kraus was a typical proponent of the idea that one can learn various things about the geometrical and physical behavior of bodies through simple "intuition." Anything that contradicted this intuition he considered absurd. Among these absurdities he included Einstein's assertion that Euclid's geometry, which we all learned in school, might not be strictly correct. Since in Kraus's opinion the truths of ordinary geometry must be clear to every normal person, it was a puzzle to him how a person like Einstein could believe the opposite. His wife reminded me not to speak to him about Einstein's theory. She said that he often spoke about it in his sleep and he got excited over the idea that there were people who could "believe what is absurd." It was tormenting for him to think that such a thing was possible.

This philosopher was the chief speaker against Einstein. I presided at this discussion and endeavored to direct it in halfway quiet paths. A number of people now appeared who wanted to take advantage of an opportunity that would probably never present itself to them again. They could now fling the opinions that they had formed privately directly at the famous Einstein; he was compelled to listen to them. As a result several comical things occurred. Thus a professor of mechanical engineering at the Institute of Technology made some remarks that were false, but sounded rather reasonable. After the lecture Einstein said to me: "That laborer spoke naïvely, but not in an entirely foolish way." When I replied that he was not a laborer, but a professor of engineering, he said: "In that case it was too naïve."

On the following day Einstein was to depart, but by early forenoon the news had already spread that Einstein was staying at the Physics Laboratory and many people hurried to speak to him. I had great difficulty in arranging a relatively quiet departure. For instance, a young man had brought a large manuscript. On the basis of Einstein's equation $E = mc^2$ he wanted to use the energy contained within the atom for the production of frightful explosives, and he had invented a kind of machine that could not possibly function. He told me that he had awaited this moment for years and in any case wanted to speak to Einstein personally. I finally prevailed upon Einstein to receive him. There was but little time left and Einstein said to him: "Calm yourself. You haven't lost anything if I don't discuss your work with you in detail. Its foolishness is evident at first glance. You cannot learn any more from a longer discussion." Einstein had already read about a hundred such "inventions." But twenty-five years later, in 1945, the "real thing" exploded at Hiroshima.

---

3. *Austria*

From Prague Einstein went to Vienna, where he also had to give a lecture. The Vienna of this postwar period was completely different from the city that Einstein had visited in 1913. It was now no longer the capital of a great empire, but only that of a little republic.

Among Einstein's acquaintances, too, changes were noticeable. His friend Friedrich Adler had become a public figure. During the war, when the Austrian government had refused to convene the parliament and to submit its course of action to the judgment of the people's representatives, Friedrich Adler, imbued with a fanatical desire to achieve what he considered just, had shot the head of the government during a dinner in a fashionable hotel.

Adler was arrested and condemned to death, but the Emperor commuted his sentence to life imprisonment, since Adler's father, although leader of the Socialists, was a man highly regarded in government circles. The hypothesis was set up that Friedrich Adler was not in his right mind when he committed the assassination. This assumption made it easier to commute his sentence, but the investigation of his mental state was remarkable. While in prison, Adler had written a work on Einstein's theory of relativity; he believed that he was able to present cogent arguments against it. This manuscript was sent by the court to expert psychiatrists and physicists. They were to determine whether any conclusion could be drawn from it that the author was mentally deranged. In this way I received a copy of the manuscript. The experts, especially the physicists, were placed in a very difficult situation. Adler's father and family desired that this work should be made the basis for the opinion that Adler was mentally deranged. But this would necessarily be highly insulting to the author, since he believed that he had accomplished an excellent scientific achievement. Moreover, speaking objectively, there was nothing in any way abnormal

about it except that his arguments were wrong. I imagine, how-·
ever, that he owed the commutation of his sentence rather to
the prestige of his father and the inclination of the Imperial
government to compromises than to the madness of his arguments against the theory of relativity.

In Vienna Einstein lived with the well-known physicist Felix
Ehrenhaft, who in his entire mode of working was the diametrical opposite of Einstein, but whom Einstein occasionally found
congenial for this very reason. Einstein was always interested in
determining how much could be deduced from a few fundamental principles. The greater the extent to which natural phenomena could be fitted into a simple pattern, the more interesting they were for him. Ehrenhaft, however, was a man of the
direct experiment. He believed only what he saw, and constantly
found isolated phenomena that did not fit into the grand
scheme. For this reason he was frequently regarded with dis-
dain, especially by persons who accepted the general scheme
as an article of faith. A man like Einstein, who had himself
brought these general principles to life, always felt mysteriously
attracted whenever he heard of irregularities. Even though he
did not believe that they existed, yet he suspected that there
might be the germs of new knowledge in these observations.

Ehrenhaft's wife was a remarkable figure among the women
of Vienna. She was herself a physicist and an outstanding or-
ganizer of education for girls in Austria. She was astonished
when Einstein arrived with only one white collar. She asked
him: "Perhaps you have forgotten something at home?" He
replied: "By no means; this is all that I need." As a good housewife she sent one of the two pair of trousers that he had brought
with him to be pressed by a tailor. But to her consternation she
noticed at the lecture that he had put on the unpressed pair.
Mrs. Ehrenhaft also thought that he had left his bedroom slippers at home and bought him a new pair. When she met him
in the hall before breakfast, she noticed that he was barefooted.
She inquired whether he had not seen the slippers in his room.
"They are entirely unnecessary ballast," was his reply. He did
not like shoes at all, and at home when he really wanted to relax he could often be seen in his stocking feet, sometimes even
when he had visitors who were not very formal.

During his stay Einstein also came in contact with the two
intellectual currents of Vienna that have most strongly influenced the intellectual life of our time: Siegmund Freud's psy-

choanalysis, and the positivistic tradition of Ernst Mach. Einstein called on Josef Breuer, a doctor who together with Freud had published the first paper on the psychological causes of hysterical paralyses, and the engineer and writer Popper-Lynkeus, the nearest friend of Ernst Mach, who once remarked that at first Popper-Lynkeus had been the only one to understand his ideas. At this time Popper-Lynkeus was already eighty years old and confined to his sofa, but intellectually he was still very alert and always eager to meet new and interesting people. He had worked out a project for the abolition of economic misery in Germany through the introduction of a general labor service. This plan was put into practice later in a distorted way by Adolf Hitler. It was a great occasion for Popper when he met Einstein who had become the true heir of Mach's ideas in the field of physics.

Einstein's lecture, which was given in an enormous concert hall before an audience of some three thousand people, was probably the first lecture of this kind that he had given. Even more than in Prague the public was in a remarkably excited state, the kind of mental state in which it no longer matters what one understands so long as one is in the immediate neighborhood of a place where miracles are happening.

---

4. *Invitation to the United States*

After Einstein's return to Berlin he was more than ever a center of general attention. Just as formerly the German professor who forgets his umbrella, the hunter who buys a hare at the butcher's shop, or the old maid looking for a man had appeared repeatedly in the German comic journals, now the name Einstein became a generic name for anyone who writes something incomprehensible and is admired on this account. Especially the word "relative" stimulated people to the most trivial jokes. In part they were malicious, trying in some way to connect Einstein's theory with the efforts of victorious France to squeeze as large reparations as possible out of Germany. The German government always tried to show that the country was completely impoverished, while the French doubted this. Thus a German comic journal represented Einstein in conversation with the French President Millerand, who was a vigorous advocate of the "Make Germany pay" policy. Millerand says to Einstein: "Can't you persuade the simple-minded Boche that even with an absolute deficit of 67,000,000,000 marks he is still *relatively* well off?"

Einstein, however, paid as little attention as possible to all these political and personal vexations and endeavored rather to dispel scientific and philosophical misunderstandings of his theories. Many people considered particularly absurd the assertions of Einstein's theory that Euclidean geometry is invalid in a gravitational field, that space is curved, and perhaps even finite. This was because everyone had learned in school that the postulates of geometry are absolutely correct, since they are not based on experience, which is fallible, but rather on infallible pure thought or on still more infallible "intuitive perception."

In a lecture delivered at the Prussian Academy in January 1921 Einstein clarified the relation between "Geometry and Experience." He said there: "In so far as geometry is certain, it says nothing about the actual world, and in so far as it says something about our experience, it is uncertain." He made a sharp distinction: On the one side is mathematical geometry, which deals only with the conclusions that can be drawn from certain assumptions without discussing the truth of these assumptions. In it everything is certain. Alongside it is a physical geometry, which Einstein used in his theory of gravitation. It deals with the results of measurements on physical bodies and is a part of physics, like mechanics. Similarly it is just as certain or uncertain as the former. This lecture through its clear formulations brought order into a field where confusion often prevailed, and in some instances still prevails even among mathematicians and physicists. Since then Einstein's formulations have been cited as the clearest and best, even by philosophers.

But while Einstein was working on this lecture, thoughts of another kind were also passing through his head. A short time before, he had received an invitation from Weizmann, the leader of the Zionist movement, to accompany him on a trip to the United States.

At a time when but few German scientists and very few German Jews had any intimation of the coming Nazi revolution in Germany, it was already fairly evident to Einstein that conditions were developing there that could become very unpleasant for him. He sensed the activities of the group growing beneath the surface that were later to come to power as the National Socialist Party. Indeed, Einstein was one of the first to feel the impact of this movement. When Einstein gave his lecture in Prague, he spoke to me about these apprehensions. At that time he thought that he would not like to remain in Germany longer than another ten years. It was then 1921. His estimate was too conservative by only two years.

The purpose of the trip that Weizmann planned was to obtain help in America for the establishment of a Jewish national home in Palestine, and in particular for the Hebrew University to be founded there. Since the American Jews were considered the wealthiest in the world, these aims could be accomplished only with their financial assistance. Weizmann laid great value on this teamwork. He hoped that Einstein's scientific fame would encourage American Jews to contribute to a noble cause. Einstein was now in a position to place his prestige at the disposal of the Zionist movement for these purposes, which he considered as having a very great educational significance for the Jews. After considering the matter for only a few days Einstein accepted the invitation.

He was motivated chiefly by the desire not only to be active as a pure scientist, but also to contribute something to the welfare of persecuted human beings. He was also impelled by the desire to see America with his own eyes and to become acquainted with the life in this new world. He felt that it would be worth while for him to know something about the great country on the other side of the Atlantic whose tradition of democracy and tolerance had always struck a sympathetic chord within him.

5. *Reception by the American People*

The arrival of Einstein and his wife in New York Harbor was accompanied by demonstrations of enthusiasm such as had probably never before been seen at the arrival of a scientist, especially not of a scientist whose field is mathematical physics. Reporters and cameramen in large numbers rushed aboard ship to photograph him or to ask him various questions.

Facing the cameras was the easiest of these ordeals. After it was over, Einstein said: "I feel like a prima donna." He also replied with a fine sense of humor to the questions that the reporters put to him. As a matter of fact, he was used to strange questions and had already developed a certain technique for answering such questions as cannot be answered rationally. On such occasions he usually said something that was not a direct answer to the question, but was still rather interesting, and which when printed conveyed to the reader a reasonable idea or at least gave him something to laugh about. Einstein was never a killjoy.

The interrogators were chiefly interested in three things. The first question was the most difficult: "How could one explain the content of the relativity theory in a few sentences?" It was probably impossible to answer this question, but it had already been put to Einstein so many times that he had prepared an answer in advance. He said: "If you will not take the answer too seriously and consider it only as a kind of joke, then I can explain it as follows. It was formerly believed that if all material things disappeared out of the universe, time and space would be left. According to the relativity theory, however, time and space disappear together with the things."

The second question was very "urgent": "Is it true that only twelve people in the world understand the theory of relativity?" Einstein denied that he had ever made such an assertion. He thought that every physicist who studied the theory could readily understand it, and that his students in Berlin all understood it. Nevertheless, this last assertion of Einstein's was certainly too optimistic.

The third question, on the other hand, was a very delicate one: the reporters asked Einstein to explain the existence of such mass enthusiasm for an abstract theory that is so hard to understand. Einstein answered with a joke. He suggested that it was a problem for psychopathological investigation to determine why people who are otherwise quite uninterested in scientific problems should suddenly become madly enthusiastic over the relativity theory and want to greet Einstein on his arrival. One of the reporters asked him whether it might not be due to the circumstance that the theory has something to do with the universe, and the universe in turn with religion. Einstein replied that it was quite possible. But in his endeavor not to permit the rise of any exaggerated opinions regarding the general significance of his theory for the great majority, he said: "But it will not change the concept of the man in the street." He explained that the only significance of the theory was that it derived from simple principles certain natural phenomena that were formerly derived from complicated principles. This is naturally important for philosophers, but hardly for the man in the street.

After this rather abstract discussion the desire to ask questions subsided somewhat and Einstein was able to close with the words: "Well, gentlemen, I hope I have passed my examination." Then, to get an element of human interest, Mrs. Einstein was asked whether she also understood the theory. "Oh no," she answered in a friendly but somewhat amazed tone, "although he has explained it to me so many times; but it is not necessary for my happiness."

Finally Mr. and Mrs. Einstein could go ashore. Einstein passed through the enormous crowd of onlookers, a brier pipe in one hand, a violin case in the other. Now he no longer appeared to the crowd as a mythical harbinger of a new system of the universe, as the man who had revolutionized space and time, but rather as a friendly musician who arrived for a concert in New York smoking his pipe.

The enthusiasm manifested by the general public on Einstein's arrival in New York is an event in the cultural history of the twentieth century. There was no single cause for this phenomenon. First of all, there was the general interest in the theory of relativity, which in itself was something astonishing. A second factor was the recognition that Einstein had received in England two years earlier, after the observation of the solar eclipse had confirmed his theory. Finally, there was something romantic about his present trip. He came not only as a scientist, but also to fulfill a political mission that was not a matter of ordinary politics but was itself surrounded with an aura of romanticism. His visit to America was his contribution to the movement whose purpose it was to enable the Jewish people to return to its homeland after having wandered over the world for two thousand years. For the Jews who felt that to a greater or lesser degree they were strangers everywhere in the world, these were happy tidings; and to every person in America it recalled the Holy Land and the legend of the Wandering Jew, thus striking a strongly responsive chord and evoking profound sympathies in many Christians.

Einstein took the whole matter very calmly. Nevertheless, he was amazed that so many people could be interested in things about which he had pondered in silence and which he had thought would probably always remain limited to a small group. Einstein's enemies often claimed that this enthusiasm had been manufactured by the press. This assertion, however, is as trivial as it is erroneous. Newspapers are constantly publicizing all sorts of things; they succeed in arousing enthusiasm for football games and movie stars; but no newspaper publicity could ever produce such enthusiasm for a mathematical physicist, even though all kinds of scientists have been publicized by the press. The reasons for this success must already have been present in the situation itself, in the unique coincidence of Einstein's achievements, his personality, and the intellectual needs of his age at the moment. When I once asked Einstein what emotions were aroused in him when he saw himself honored in this way, he replied: "The impression cannot be very elevating when I remember that a victorious boxer is received with still greater enthusiasm."

He himself was always inclined to see the causes of this phenomenon rather in the disposition of the public than in his own person. Thus he sometimes jokingly remarked: "The ladies in New York want to have a new style every year — this year the fashion is relativity."

Nevertheless, if one considers the matter realistically and dispassionately, one must ask with amazement: how was it that a mathematical physicist became as popular as a boxer? Seen objectively, this was indeed a good indication of popular taste in New York. It may have been simply a desire for sensation, but if so, why was this popular interest centered on Einstein?

Some persons regarded it as an indication of the high cultural level of the American people. This was the view of the editor of the best popular scientific journal: "No European populace would welcome a distinguished scientist with such enthusiasm. America doesn't boast a leisure class that takes conventional interest in science and philosophy. But the figures of reading and educational efforts justify the belief that Einstein should have taken the honors bestowed upon him at their face value, as evidence of a profound popular interest in the field where he has so few peers."

Perhaps it will seem strange to some people, but the truth is that Einstein never worried his brain very much regarding the reasons for this interest. His attitude toward the world around him was always to some extent that of an onlooker at a performance. He was accustomed to believe that many things are incomprehensible, and human behavior was not one of the things in which he was most interested. As a normal, natural person he was happy when he was received with friendliness and goodwill, without inquiring too much into the reasons for such kindliness. He was never inclined to have too high an opinion of the goodwill of the public or to make any concessions to it. His utterances were never calculated to evoke cheap applause. In later years he was well aware that many persons paid a great deal of attention to everything he said, and that it was important to utilize this power over people for educational purposes. For this reason in interviews with newspapermen he often said things that were not very pleasant or very comprehensible to the readers of these newspapers. His idea was that when an opportunity presents itself, good seed should be sown. Somewhere some of it will sprout.

Einstein had put himself at the disposal of the Zionist leaders with the idea that his presence would help their propaganda for the Jewish National Fund and especially the collection of contributions for the university in Jerusalem. At the meetings that were organized for these purposes in many places throughout the United States, he sat near Weizmann, generally in silence, sometimes speaking a few words in support of him. He sincerely wanted to be a faithful member of the movement for the rebirth of the Jewish people. At one meeting he spoke after Weizmann, quite as if he was a member of the rank and file who wanted no personal prominence but wished only to serve the cause. He said: "Your leader Dr. Weizmann has spoken and he has spoken very well for us all. Follow him and you will do well. That is all I have to say." This sounds almost as if it was spoken in the spirit of the leadership principle. In some respects it was probably a relief for Einstein, who always stood alone, to feel himself a member of a popular movement rooted in the broad masses. But this feeling was always of only short duration. Inevitably there soon reappeared his aversion to everything that ties him to a party, even though in certain respects it may have been congenial to him.

Einstein and Weizmann were regarded by all official personalities in America as authorized representatives of the Jewish people and greeted as such. President Harding wrote in a letter to a meeting at which Einstein and Weizmann spoke: "Representing as they do leadership in two different realms their visit must remind people of the great services that the Jewish race has rendered to humanity."

Similarly Mayor Hylan of New York in welcoming them at the City Hall addressed them as the representatives of their people, saying: "May I say that in New York we point with pride to the courage and fidelity of our Jewish population demonstrated in the World War."

The Jewish population of America itself regarded Einstein's

visit as the visit of a spiritual leader, which filled them with pride and joy. The Jews felt that their prestige among their fellow citizens was raised by the fact that a man of Einstein's generally recognized intellectual greatness publicly acknowledged his membership in the Jewish community and made their interests his own. When Einstein arrived with Weizmann in Cleveland, all the Jewish businessmen closed their establishments so as to be able to march in the parade that accompanied Einstein from the Station to the City Hall. When Einstein and Weizmann addressed Zionist meetings, it seemed almost as if the political and spiritual heads of the Jewish people were appearing together.

These appearances in the service of the organization that represented some of his political and cultural aims were interspersed with lectures on his scientific theories. Sometimes he appeared in a most informal manner. Thus he visited Professor Kasner's class at Columbia University just as he was explaining the theory of relativity to his students. Einstein congratulated Kasner on the comprehensible manner in which he did this, and then spoke to the students himself for about twenty minutes.

Later he addressed the students and faculty of Columbia University and was greeted by the outstanding physicist Professor Michael Pupin. This remarkable man, once a Serbian shepherd, had become one of the leading inventors and scientists in the world, and through his understanding of electrical phenomena the first transatlantic telephone cable had been made possible. He regarded all theories with the dispassionateness of a laboratory worker, but, unlike so many others, greeted Einstein not as a person who invented absurd and sensational things but as the "discoverer of a theory which is an evolution and not a revolution of the science of dynamics."

At this time Einstein always lectured in German, because he did not yet have full mastery of English. On May 9 he received an honorary degree from Princeton University. President Hibben of the university lauded him in a German address: "We salute the new Columbus of science voyaging through the strange seas of thought alone." Later Einstein gave several lectures at Princeton, in which he presented a comprehensive survey of the theory of relativity.

Einstein was regarded, however, not only as a representative of the Jewish people. Since he had left his work at the Berlin Academy to come to America, and because he spoke German everywhere, he was also considered a representative of German science. In view of the fact that it was not long after the war, this aroused hostile reactions in some quarters.

Sometimes semi-comical occurrences took place when political attacks were directed against him and no one knew whether he was attacked as a Jew or as a German. An episode of this kind occurred when Fiorello H. La Guardia, then president of the Board of Aldermen of New York City, proposed that Einstein be given the "freedom of the City of New York." All the aldermen were in favor of this resolution but one, who declared "that until yesterday he had never heard of Einstein." He asked to be enlightened, but nobody offered to explain the theory of relativity. But the Jews and the Germans did not believe in the naïveté of Einstein's opponent. He was accused of partly anti-Semitic, partly anti-German opinions. He defended his action on patriotic grounds: he wanted to spare his beloved native city from the possibility of becoming a scientific and national laughing stock. He said at the session: "In 1909 the key of the city was unfortunately given to Dr. Cook, who pretended to have discovered the North Pole." Perhaps, he suggested, Einstein had not really discovered the theory of relativity. Besides, he continued, "I have been assured that Professor Einstein was born in Germany and was taken to Switzerland but returned to Germany prior to the war. He is consequently a citizen of Germany, of an enemy country, and might be regarded as an enemy alien."

Everyone was so interested in Einstein's theory and its meaning that Congressman J. J. Kindred of New York requested the Speaker of the House of Representatives for permission to publish a popular presentation of the relativity theory in the *Congressional Record*. Representative David Walsh of Massachusetts had his doubts about permitting anything to appear in the *Record* that had nothing to do with the activities of Congress and that in addition seemed to be incomprehensible.

"Well, Mr. Speaker," said Representative Walsh, "ordinarily we confine matters that care to appear in the *Congressional Record* to things that one of average intelligence can understand. Does the gentleman from New York expect to get the subject in such shape that we can understand the theory?" Kindred answered: "I have been earnestly busy with this theory for three weeks and am beginning to see some light." But then Representative Walsh asked him: "What legislation will it bear upon?" To this Representative Kindred could only reply: "It ι.ιay bear upon the legislation of the future as to general relations with the cosmos."

At the time when Einstein was in the United States, a statement by the great inventor Thomas Edison created quite a furore throughout the country. He denied the value of college education and asserted that education should be directed toward learning relevant facts. He worked out a questionnaire containing questions that he thought were relevant for practical people, and suggested that tests be made, which would show that most college graduates were unable to answer these questions.

While Einstein was in Boston, staying at the Hotel Copley Plaza, he was given a copy of Edison's questionnaire to see whether he could answer the questions. As soon as he read the question: "What is the speed of sound?" he said: "I don't know. I don't burden my memory with such facts that I can easily find in any textbook." Nor did he agree with Edison's opinion on the uselessness of college education. He remarked: "It is not so very important for a person to learn facts. For that he does not really need a college. He can learn them from books. The value of an education in a liberal arts college is not the learning of many facts but the training of the mind to think something that cannot be learned from textbooks." For this reason, according to Einstein, there can be no doubt of the value of a general college education even in our time.

Einstein was often mentioned together with Edison, both being honored as the outstanding representatives of physical science. Edison was to the technical application of physics what Einstein was to its theoretical foundation.

Einstein also visited the physics laboratories of the oldest university in the United States, Harvard University. Professor Theodore Lyman, famous for his optical investigations, informed him about the work that was being done there. Lyman had the feeling that after the many meetings at which Einstein had been used as an instrument of political propaganda, even though for a purpose with which he was entirely in sympathy, he now could breathe freely, being again in the atmosphere of a physics laboratory and able to immerse himself in the problems of nature. Most visitors to a laboratory rapidly pass by the experimental arrangements and listen only half-heartedly to the explanations of the students. Einstein, however, did not remain satisfied with a superficial "That is very interesting," or some similar polite remark; instead, he allowed several students to give him detailed explanations of the problems on which they were working. Furthermore, he actually thought about these problems, and some students received advice from him that was helpful in their research. Such absorption during a strenuous journey is only possible for a man possessing two qualities that are rarely found together: first, an unusual ability to familiarize himself rapidly with an unfamiliar problem; and secondly, the capacity to enjoy helping someone who is doing scientific research.

There is no doubt that Einstein made his first trip to America not only in the service of science and of the future university in Jerusalem, but also because he was particularly interested in becoming acquainted with life on this continent, which was new to him. This first trip, however, was not very favorable for the achievement of this purpose. The entire journey proceeded at a whirlwind pace, leaving him no time for any quiet reflection. As a result the impressions that Einstein received on his first visit to the United States could only be very superficial ones that struck one at first glance. In the first place he was impressed by American youth, with its fresh energetic urge to acquire knowledge and to do research. He once said: "Much is to be expected from American youth: a pipe as yet unsmoked, young and fresh." Then there was the impression of the many peoples that had settled America, which despite their different origins lived together in peace under a tolerant democratic regime. He remarked in particular about New York City: "I like the restaurants with the color of the nations in the air. Each has its own atmosphere. It is like a zoological garden of nationalities, where you go from one to another." He was also struck by the role of women in American life, observing that it was much greater than the part played by women in European life.

Efforts were made to enlist Einstein's interest in campaigns to restrict the use of tobacco and Sunday amusements. In such matters, however, Einstein did not favor any excessive restrictions on individual liberty. He was much too natural a person not to have recognized the importance of the innocent pleasures of daily life. He did not have any faith in cut and dried schemes for making people happy by dictating to them what they are to regard as work and what as play. Replying to a man who had requested his opinion on the matter of Sunday rest, he said: "Men must have rest, yes. But what is rest? You cannot make a law and tell people how to do it. Some people have rest when they lie down and go to sleep. Others have rest when they are wide awake and are stimulated. Some must work or write or go to amusements to find rest. If you pass a law to show all people how to rest, that means you make everybody alike. But everybody is not alike."

Einstein, who devoted his entire life to the discovery of physical laws that could be derived from a few general principles, was not of the opinion that life could be regulated according to a few abstract principles. He was always more inclined to rely on the natural instincts. As a passionate smoker he also remarked on that occasion: "If you take tobacco and everything else away, what have you left? I'll stick to my pipe."

He often had experiences that made it difficult for him to maintain his equanimity. His naïve joy in simple pleasures such as smoking certainly helped him in these situations. Ascetic instincts were foreign to him.

---

6. *England*

The report of the English astronomers to the Royal Society in London in 1919 had laid the foundation for Einstein's world fame. But Einstein himself had not yet been in London. In 1919, in the postwar atmosphere of hostility to Germany, it had indeed been possible to recognize the theory of a German, but not to honor a German personally. Lord Haldane, who had always worked for the improvement of Anglo-German relations, had been in Berlin shortly before Einstein's arrival there, but had received a cool reception from the Kaiser. Immediately after the war and Germany's defeat, however, Haldane again began to build up new cultural relations with Germany. Einstein seemed to him to be a person who could serve as the thin end of a wedge with which one could penetrate the mass of hostility and prejudice. Many favorable factors seemed to be present: the great acclaim that Einstein's prediction of the result of the solar eclipse expeditions had produced; the opportunity for a great achievement that had thus been presented to English science; and finally, also, the favorable circumstance that Einstein did not belong to the hated kind of German; indeed, if one so desired, he could be regarded as a non-German. It thus seemed almost as if Einstein had been specially created to act as an intermediary. In addition, for Lord Haldane there was a very important personal factor. He was one of those English statesmen whose hobby was science combined with philosophical speculation. Haldane had set himself the problem of how, despite the skepticism that had become prevalent in religion, morals, politics, and even science as a result of the disappointments of the postwar period, one could still retain an objective conception of truth. In his book *The Reign of Relativity,* published in 1921, he pointed out that the views that skeptics regard as different are actually only different aspects of the same truth and that therefore a single objective truth exists. Or, in Haldane's own words:

"The test of truth may have to be adequacy in a fuller form, a form which is concerned not only with the result of measurement with the balance or rule, but with value, that cannot be so measured and that depends on other orders of thinking. What is truth from one standpoint may not of necessity stand for truth from another. Relativity, depending on the standard used, may intrude itself in varying forms. . . . It may, therefore, be stated generally that an idea is true when it is adequate, and only completely adequate when it is from every point of view true. Each form of test that is applicable must be satisfied in the conception of perfect adequacy; for otherwise we can have only truth that is relative to particular standpoints."

This philosophy found its practical application in a training for tolerance toward one's fellow man and in the struggle against the overestimation of political doctrines. In Einstein's theory Haldane saw a special example of his own philosophy. He believed that the physical theory of relativity would invest his philosophy of relativity with greater certainty and an increased brilliance. Consequently Haldane endeavored to induce Einstein to stop over in England for several days on his return from America, to give several lectures there, and personally to meet scientists and people in public life.

Not only were there political difficulties in the way of such personal contacts, but the entire mental attitude of the English physicists was not such as to make them very enthusiastic about a theory like that of relativity. English science was always much more intent upon the direct connection between experiment and theory. A connection that consisted in such long chains of thought as in Einstein's theory often appeared to the English physicist to be a philosophical phantasm — too much theory for so few facts. In England philosophers, astronomers, mathematicians, even theologians and politicians were passionately interested in the theory, but the physicists themselves were still rather cool to "Relativity" as a basic concept.

Lord Haldane presided at Einstein's lecture at King's College.

He introduced the lecture by saying that it had been an extremely moving moment for him when Einstein laid a wreath
on Newton's grave in Westminster Abbey. "For," Haldane told
the audience, "what Newton was to the eighteenth century, Einstein is for the twentieth."

In Haldane's house, where Einstein lived, he met many
famous Englishmen, like Lloyd George, Bernard Shaw and
A. N. Whitehead, the mathematician and philosopher, who had
so vividly sensed the historic significance of the session of the
Royal Society at which the result of the solar eclipse had been
announced. Whitehead had long discussions with Einstein and
repeatedly attempted to convince him that on metaphysical
grounds the attempt must be made to get along without the assumption of a curvature of space. Einstein, however, was not inclined to give up a theory, against which neither logical nor
experimental reasons could be cited, nor considerations of simplicity and beauty. Whitehead's metaphysics did not seem quite
plausible to him.

The Archbishop of Canterbury, the head of the Anglican
Church, was especially desirous of meeting Einstein. Lord Haldane, who called attention everywhere to the philosophical significance of the relativity theory, had told him that this theory
also has important consequences for theology and that as head
of the Anglican Church it was his duty to become acquainted
with it. Shortly thereafter, at the Athenæum Club, a friend of
the Archbiship met J. J. Thomson, the physicist and president
of the Royal Society, and requested his help in a very important
matter. "The Archbishop, who is the most conscientious of men,
has procured several books on the subject of relativity and has
been trying to read them and they have driven him to what, it
is not too much to say, is a state of intellectual desperation. I
have read several of these myself and have drawn up a memo-
randum which I thought might be of service to him."

Thomson was surprised by these difficulties and said he did
not think that the relativity theory was so closely connected with
religion that the Archbishop had to know something about it.
Nevertheless, the conscientious head of the church was not satisfied, and when Einstein came to London and Lord Haldane arranged a dinner the Archbishop asked for an invitation. He was
placed as Einstein's neighbour and was able to hear whether
Haldane was right in his assertion that the theory of relativity is important for theology, or Thomson, who disputed it.
At dinner the Archbishop asked bluntly "what effect relativity

would have on religion." Einstein replied briefly and to the point: "None. Relativity is a purely scientific matter and has nothing to do with religion."

7. *Einstein Tower and the Rathenau Murder*

In June 1921, after visiting the United States and England, Einstein returned to Berlin. The honors that he had received abroad had their effect in Germany. Well-meaning persons who were not really interested in science tried in every possible way to learn about Einstein's theories without having to exert themselves mentally. As a result some people profited from this boom by convincing others that they could teach the relativity theory. At this time, for instance, a so-called "Einstein film" was shown in movie theaters that was supposed to teach the theory painlessly. First it showed a student listening to a boring lecture by a dull professor and sighing: "How much longer will this lecture last? Still another quarter of an hour?" Then it showed the same student sitting on a bench in a garden with a pretty girl, complaining: "I can only stay fifteen minutes more." This was supposed to teach the "relativity of time" to the public. As we have seen, this has nothing to do with Einstein's theory. Such popularizations, which distorted and made the theory trivial, vexed Einstein more than the attacks on him.

In Berlin people also amused themselves with anecdotes from England. For instance, an imaginary conversation between Einstein and Bernard Shaw was described in which the skeptical author asks: "Tell me, my dear Einstein, do you really understand what you wrote?" And Einstein, smiling at him, replies: "As much as you understand your things, dear Bernard."

At this time the desire for a short, easily comprehensible presentation of the theory of relativity led an American who was living in Paris and who had been impressed by the London reports on the solar eclipse expeditions to offer a prize of five thousand dollars for the best essay on Einstein's theory in not more than three thousand words. Attracted by the remuneration of five dollars for three words, many entered the contest, and, indeed, it was rather difficult to find judges, since everyone acquainted with the subject preferred to enter the contest. Einstein remarked playfully: "I am the only one in my entire circle of friends who is not participating. I don't believe I have the ability to accomplish the task." On June 21, 1921, out of the three hundred essays submitted, the prize was awarded to a sixty-one-year-old Irishman, a native of Dublin, who like Einstein had been employed in a patent office for a long time (in London), and who was a dilettante in physics. It can hardly be said that his essay was better than those of his competitors, nor did it have any further influence in spreading an understanding of the relativity theory. The public remembered only the fact that someone had been able to earn five thousand dollars by means of it, and concluded that it must therefore be worth the effort of studying it.

In the fall of 1921 an important step was taken to investigate another of the astronomical conclusions of Einstein's theory. Dr. Bosch, the director of I. G. Farben, the greatest chemical concern in Germany, that was outstanding in the production of synthetic dyes, medicaments, and explosives, donated a large sum of money for the erection at Potsdam of an institute to be connected with the Astrophysical Observatory, where the color composition of solar rays could be investigated with great precision. It will be recalled that from his theory of gravitation Einstein had predicted that the color of light coming to us from the stars depends on the intensity of the gravitational field through which the rays pass. This prediction was to be verified by exact observations.

The astronomer Erwin Finlay-Freundlich was appointed director of this institute. The laboratory was built in the form of a tower and the architectural design was in the characteristic modern Berlin style of that period, so that the result was a cross between a New York skyscraper and an Egyptian pyramid. The tower became generally known as the *Einstein Tower*. Its appearance alone was enough to excite the ire of the nationalistic groups who preferred a style more reminiscent of German medieval models or at least of classical antiquity.

Through a strange concatenation of circumstances, the Einstein Tower was under the control of Friedrich Wilhelm Ludendorff, a brother of the famous general Ludendorff who for a long time collaborated with Adolf Hitler. At that time the astronomer Ludendorff still permitted the investigation of solar light on the basis of Einstein's theory. He satisfied his nationalistic sentiments by endeavoring to prove that Copernicus was a German and not a Pole, even though his monument stood in Warsaw.

On June 24, 1922 Walther Rathenau, then minister of foreign

affairs, was murdered by several fanatical students. This murder revealed the preparations for the revolution from the Right, and even those who consciously or unconsciously ignored the background of this crime were compelled to take a more serious view of the matter. The effect on Einstein was more intense than on the general masses because through his insight and instinctive perception it had become clear to him that sincere allegiance to the German Republic was confined to a small group; beneath it yawned a hate-filled abyss.

Einstein had been acquainted with Rathenau and had liked this man whose breadth of vision was so rare among German politicians. Rathenau, a descendant of a rich Jewish family in Berlin, had been the motive force behind the planned economy in Germany during the war. After the proclamation of the Republic, Rathenau had played an important part as an economic adviser to the government, and through his international reputation he had been able to perform various services in aid of its foreign policy. During the government of the Catholic Chancellor Wirth, Rathenau had accepted the position of Foreign Minister and by concluding the Treaty of Rapallo had entered into friendly relations with Soviet Russia. This treaty served to stamp him as a "Bolshevik," and moreover, being a Jew, he had become extremely unpopular with the monarchists and the advocates of the "revolution from the Right."

The republican government ordered the day of Rathenau's burial to be observed as a day of mourning and ordered all schools and theaters to remain closed. At the universities lectures were canceled, but Philipp Lenard, the Heidelburg physicist who has already been mentioned as an opponent of Einstein, refused to obey the order. While the Socialist workers marched through the city and organized protest meetings against the murderers and their reactionary supporters, Lenard demonstratively gave his regular lecture. A number of students who sympathized with the assassins listened to him with enthusiasm. A group of workers passing by the building saw the lecture going on and, regarding this as a demonstration in favor of the murderers, entered the building and dragged Lenard out with them. As they passed over the Neckar River some of them attempted to throw Lenard into the water, but the moderates prevented them and turned him over to the police, who immediately released him.

In the eyes of all Germany these events linked the fight against Einstein's theory with the struggle against the republican regime. Rumors began to spread that in the search for Rathenau's murderers a list had been found containing the names of other persons who were to be the future victims of the same group, and the list was supposed to contain Einstein's name. The police denied the rumors, but a feeling of uneasiness concerning Einstein's person began to spread. Einstein himself, with his belief in the inevitable in the universe, had no inclination for superstitious presentiments and fears and was not affected. But the reaction on those about him was all the greater.

Every year the annual meeting of German scientists and physicians took place in September. This year a special celebration was planned, as it was the centennial meeting. It was to be held in Leipzig. Because during the past few years he had contributed most to increase the prestige of German science throughout the world, Einstein was invited as keynote speaker to emphasize the special character of the occasion. He would have liked to accept the invitation, but in the troubled and uneasy atmosphere that prevailed after the Rathenau murder he did not wish to make any public appearances and declined to participate. Nevertheless, the executive committee of the society insisted on presenting lectures by other scientists on the significance of Einstein's theory, one by the physicist Max von Laue, the other by the philosopher Moritz Schlick.

Because of the spreading sentiment in favor of violence, and as a result of his own adventure as well, Lenard felt himself compelled and encouraged to protest against the meeting of the German scientists in Leipzig. In his opinion, by arranging for lectures on Einstein's work, the society of German scientists was carrying on propaganda against the revolutionists of the Right and on behalf of the group to which the "justly" murdered Rathenau belonged. Lenard assembled a group of people who drew up a protest against the meeting of the German scientists, which was sent to all the newspapers and distributed in Leipzig at the doors of the lecture halls.

Lenard did not succeed, however, in inducing any of the creative German physicists to sign his protest. Once again it was only the same three types of persons who had taken part in the meetings at the Berlin Philharmonie.

8. *France*

Einstein's travels had contributed somewhat to improve the relations between the scientists of Germany and those of America and England. This was agreeable to the government of the German Republic and to the German scientists, but was very annoying to all those groups that endeavored to maintain the idea that in western Europe the Germans were looked down upon as an inferior nation and there was a desire to destroy their culture. The effect of this "atrocity propaganda" was disturbed by the reports of the friendly reception granted to Einstein. It had long been discussed whether Einstein would now be bold enough to visit Paris, the capital of Germany's "mortal enemy." It had been rumored that scientific groups in France were trying to induce Einstein to make such a visit, so as to be able to discuss the new theories with him personally. They had also been greatly admired in France, but many persons had found it very difficult to understand them. Thus the French mathematician Paul Painlevé, who as Minister of War during the World War, and later as Premier and President of the French Chamber of Deputies, played a leading role in French politics, was much interested in Einstein's work, but misinterpreted it at many points and attacked it because of this misunderstanding. Later he withdrew all his objections. The great French physicist Paul Langevin, who immediately grasped the meaning of the Einsteinian theories, once remarked to me: "Painlevé studied Einstein's work very closely, but unfortunately not until after he had written about it. Perhaps he is used to this sequence from politics."

Langevin was not only a sagacious scientist but also an active participant in every enterprise intended to advance international conciliation. At the Collège de France, the highest scientific school in France, he presented a resolution to invite Einstein to come to Paris. For this purpose he proposed to use the income of an endowment that had been employed to invite other outstanding foreign scientists to lecture at this institution. The resolution was warmly supported by Painlevé. There was some opposition, however. The nationalists did not want the reception of a German scientist to arouse an impression that any diminution of their hatred was possible. With all sorts of threats they tried to induce Langevin and his friends not to extend the invitation, just as the analogous groups in Germany attempted to force Einstein to reject it. At that time, however, neither of these two groups was yet strong enough to achieve its aim. Einstein accepted the invitation extended by the Collège de France, and toward the end of March 1922 he went to Paris.

Langevin, the physicist, and the astronomer Charles Nordmann went ahead to Jeumont on the Belgian border to meet Einstein and traveled with him to Paris. During the trip they discussed the scientific and political questions connected with this visit. In the course of this conversation they asked Einstein for his opinion on the aims and influence of the Left in German political and cultural life. "Well," replied Einstein, "what is superficially called the Left is actually a multidimensional structure." Einstein already felt that the roads to the Right and to the Left might occasionally lead to the same end.

Throughout the entire trip Langevin was rather worried. Before his departure from Paris it had been rumored that groups of the "Patriotic Youth" and other nationalistic groups would gather at the station and give Einstein an unfriendly reception. Both Langevin and the French officials did not want any disturbances of this kind to spoil Einstein's visit. While on their way, Langevin asked for information about the situation in Paris. He received a telegram from the Paris police informing him that groups of excited young people were gathering at the Gare du Nord, where the trains from Belgium arrived. Since it was believed that they were the "patriots," Langevin was advised to leave the train with Einstein on a sidetrack where no one expected him. They did so, and Einstein was quite happy to be able to slip away from the train through a side entrance of the station into the street without being bothered by reporters or cameramen, and to ride in the subway to his hotel unnoticed by anyone.

At the Gare du Nord, however, a crowd of students who had gathered under the leadership of Langevin's son to give Einstein an enthusiastic reception and to prevent possible hostile demonstrations by the "patriots," waited in vain for his arrival. It was these admirers of Einstein whom the police had regarded as a hostile crowd, and it was from them that Einstein had fled.

On March 31 Einstein gave his first lecture at the College de France. Only people with tickets were admitted, the tickets having been given only to persons who were known to have an actual interest in the subject and who would not attend simply to organize a hostile demonstration. Former Premier Painlevé stood at the door himself and watched to check that only people with invitations were admitted.

Einstein spoke in the hall where great philosophers such as Ernst Renan and Henri Bergson had lectured before large audiences. Here it was easier than in England and America for him to come into contact with his audience, since he spoke French fluently and confidently, but with a slowness to which the French were unaccustomed and which together with his slight foreign accent gave his speech a certain charm and attractiveness — the charm of pensiveness combined with a trace of mystery. This slight trace of mystery contrasted with the evident effort to present everything as logically and clearly as possible, using as few technical expressions and as many metaphorical comparisons as possible. Many internationally known scholars and persons in public life attended the lecture, among them Madame Curie, the discoverer of radium, the great philosopher Henri Bergson, Prince Roland Bonaparte, and many others.

Besides this public lecture there were sessions of the philosophical and mathematical society for scientists who wanted a detailed discussion where everyone could put questions to Einstein and raise all kinds of objections. Einstein answered every question thoroughly and many misunderstandings were cleared up.

It was very strange that the Society of French Physicists did not take part officially in any of these arrangements even though many of its members naturally met Einstein. This attitude was determined chiefly by its nationalistic tendencies, which, it seems, are stronger among physicists and technicians, than among the more abstractly thinking mathematicians, astronomers, and scientific philosophers.

As in Germany, a certain resistance among the "pure" experimental physicists may also have been involved. In France there were also "pure empiricists," the kind of physicists about whom Einstein often remarked: "Everything that they learned up to the age of eighteen is believed to be experience. Whatever they hear about later is theory and speculation."

The famous Academy, which had been attacked and ridiculed for years in French literature as a center of all kinds of prejudices, likewise maintained its reputation on the occasion of Einstein's visit. There were long discussions whether Einstein should or could be invited to give a lecture. Some members maintained that it was impossible because Germany was not a member of the League of Nations. Others, in turn, thought that such an invitation would give rise to a difficult question of etiquette. Since Einstein was not a member of the Academy, he could not sit among the members, but would have to sit in the audience. Such an unhonorable seat, however, could not be offered to so famous a man. Finally thirty members of the Academy stated very bluntly, without any subtle phraseology, that if Einstein entered the room, they would immediately leave. In order to spare his French friends any unpleasantness and annoyance, Einstein himself declined to participate in a session of the Academy.

On this occasion a Paris newspaper inquired derisively: "If a German were to discover a remedy for cancer or tuberculosis, would these thirty academicians have to wait for the application of the remedy until Germany joined the League?"

The reception in Paris had shown that the need for an understanding of the modes of thought and methods of work of different peoples and individuals existed among scientists in all countries, and could be satisfied if there were a few courageous men. It also became clear that everywhere the forces of ultranationalism waited only for a suitable occasion in order to appear on the surface. In order to be able to judge these events correctly one circumstance must not be forgotten. Exactly the same groups that protested violently against the reception of Einstein because he was a German became the most zealous proponents of a policy of "collaboration" with Germany after the Nazis had seized the power there. These French "patriots" prepared the French defeat of 1940 and the German domination of the Continent.

In France just as in Germany it was evident that the attitudes of people to Einstein depended greatly on their political sympathies, since most of them made no serious effort to form an opinion about his theories. A famous historian at the Sorbonne put it as follows: "I don't understand Einstein's equations. All I know is that the Dreyfus adherents claim that he is a genius, while the Dreyfus opponents say he is an ass." Dreyfus was a captain in the French army who in 1894 had been accused of treason by anti-Jewish propagandists. The affair developed into a struggle between the Republic and its enemies, and the entire country was divided into two camps, the defenders of Dreyfus and their opponents. "And the remarkable thing," added this historian, "is that although the Dreyfus affair has long been forgotten, the same groups line up and face each other at the slightest provocation."

In Germany the republican government was attacked because

it had allowed Einstein to go to Paris and "overtures" to be made to the French; and in France the mathematicians and philosophers were attacked because they wanted to listen to one "whose people killed our sons." And when Einstein returned to Berlin and attended the first session of the Prussian Academy again, quite a few of the seats around him were empty.

9. *China, Japan, Palestine, and Spain*

After these journeys to England and France, where his stay was always bound up with political tensions and where it was really impossible to enjoy the new experiences, it was a relief for Einstein to travel to the countries of the Far East, to experience the varied impressions made upon him, and like a child at play to enjoy the variety of the world without having to consider constantly whether or not national sensibilities at home or abroad were being insulted.

Einstein arrived in Shanghai on November 15, 1922 and in Kobe, Japan, on November 20. He remained in Japan until February, when he sailed for Europe.

He was honored everywhere not only as a scientist, but also as a representative of Germany. In Shanghai he was greeted at the pier by the teachers and pupils of the German school, who sang *Deutschland, Deutschland über alles*. In Japan he was received personally by the Empress, who conversed with him in French.

When I once asked Einstein whether he had not experienced many strange things in his travels through these picturesque and exotic countries, he replied: "I have seen strange things only in my homeland — for instance, at the sessions of the Prussian Academy of Science."

The Orientals — the Hindus, the Chinese, the Japanese — with their calmness, meditativeness, and politeness, enchanted Einstein. Their liking for moderation and beauty was for him a true relaxation after the exaggerated glorifications and animosities he had experienced in his own country and its immediate neighbors.

With his preference for the music of Mozart, Bach, and the old Italian masters, Oriental music necessarily appeared very strange to him. He was unable to discover anything enjoyable in it. He was impressed, however, by the love for art, that makes

Japanese families often spend a good part of the day in the theater listening to the music, bringing their food with them and not stirring from the spot.

In certain respects it was a similar attitude when hundreds of Japanese listened patiently to Einstein's lectures without understanding even the language in which he spoke, let alone the content. One time Einstein observed that his lecture, together with the added Japanese translation, lasted more than four hours. He was shocked by this fact, because he pitied the people who listened so long and patiently to him, most of them without understanding much of what he said. When he gave his next lecture he shortened it so that it lasted only two and a half hours. While riding in the train to the next city, he noticed that his Japanese companions were whispering to each other in Japanese, looking at him, and then whispering again. Einstein began to feel uneasy, because such behavior was quite unusual in view of the politeness of the Japanese. Finally Einstein said to one of his companions: "Please tell me quite frankly if there is something amiss." Thereupon the polite Japanese answered with embarrassment: "We did not dare to say anything to you about it, but the persons who arranged the second lecture were insulted because it did not last four hours like the first one. They considered it as a slight."

On his way back, Einstein visited Palestine. For him this land was in a different category from China or Japan. Here he was unable to be simply an unparticipating observer, viewing the varied scenery as a pleasant relaxation from his work. Here he was to experience tensions that were both pleasant and unpleasant, because Einstein himself had carried on propaganda for the development of a Jewish national home in Palestine and to a certain degree felt himself responsible for it. Naturally, however, many things were not carried out as he would have desired, with the result that many people held him responsible for things with which he himself was not in sympathy. Einstein's collaboration in the development of Palestine was always directed only toward the advancement of the main goal, which he regarded as desirable. Of the concrete details of this development, only very few could be attributed to his suggestions. Consequently he was curious to see the actual appearance of what until then had been only a more or less vague dream.

As one of the most prominent advocates of Jewish colonization and as one of the outstanding personalities among the Jews throughout the world, he was received in Palestine, even more

than in other countries, as a public figure. He was invited by
the Governor of Palestine to live at his house. The Governor
at this time was Viscount Herbert Samuel, a man who had already acquired a reputation in English domestic politics. He
was himself a Jew, a fact that the English government apparently considered a particularly appropriate manifestation of its
friendly attitude toward the development of the Jewish national
home. In practice, however, things did not work out so well.
The position of a Jewish governor was particularly difficult in
the face of the growing controversies between the Jews and
Arabs. Daily he had to prove the absolute impartiality of the
English government in this conflict. Since he himself was a
Jew, it was only natural to attribute to him a certain bias in
favor of the Jews, so that he had to compensate for this by
leaning over backward in favor of the Arabs, with the result
that in the end he discriminated against the Jews. He could not
help making himself generally unpopular.

Like Lord Haldane, Viscount Herbert Samuel was one of
those English statesmen whose hobby was to occupy themselves
with science, especially with the philosophy of science. Like
Haldane he too had a strong personal interest in the relativity
theory. As regards its philosophical interpretation, Herbert Samuel's views were opposed to those of Einstein and were more
along the lines of traditional philosophy.

In the land, which was regarded more or less as a colony, an
English governor had to present an imposing front in order to
keep the "natives," including both Jews and Arabs, obedient and
respectful. When he left his palatial residence a cannon was
fired, and when he rode through the city he was accompanied
by mounted troops. Within his residence there prevailed a ceremonial formality reminiscent of the ceremonial practices at the
English court. It was necessary to arouse in the "natives" a
sense of awe in the presence of the direct representative of the
King. Einstein did not pay too much attention to all this. He
was as simple and natural as anywhere else. Mrs. Einstein, however, felt rather uneasy. She said later: "I am a simple German
housewife; I like things to be cozy and comfortable and I feel
unhappy in such a formal atmosphere. For my husband it is a
different matter; he is a famous man. When he commits a breach
of etiquette, it is said that he does so because he is a man of genius. In my case, however, it is attributed to a lack of culture."
Sometimes, to avoid the difficulties of etiquette and ceremonial,
she went to bed.

Einstein studied with the greatest interest the work of the Jews in developing an independent national life. He saw the new Jewish city of Tel-Aviv. In Europe the Jews usually belonged to only one particular class of the population; they were often persecuted by other classes, who represented the work performed by the Jews as being especially easy or particularly obnoxious. Tel-Aviv, however, was a city in which all work was done by Jews. Here they could not so easily acquire the feeling of occupying an abnormal position as an ethnic and economic group.

Nevertheless, Einstein also saw the difficulties of the Jewish situation — above all, the unsatisfactory relations with the Arabs. He was not enough of a national partisan to do what so many others did — that is, simply put the blame on the ingratitude of the Arabs and the insufficient support of the Jews by England. He demanded on the part of the Jews an effort to understand the cultural life of the Arabs and to make friends with them.

For this reason not all the Zionist groups welcomed Einstein. The extreme nationalists looked upon him just as suspiciously as the adherents of Jewish religious orthodoxy. The latter took it a little amiss that he did not consider the observance of the ancient rites important and that occasionally he even ventured a joke.

In March 1923 Einstein returned from Palestine by boat to Marseille. Thence he traveled to Spain, whose landscape and art were always a source of joy to him. Just as he had conversed with the Empress of Japan, so Einstein also had a conversation with King Alfonso XIII of Spain. Thus he not only saw strange lands and cities, but also obtained a personal impression of a class of people that usually remains unknown to scientists. Einstein, who always retained something of the curiosity of an intelligent child gathered new strength for his creative work from all these experiences. Everything seemed to him like a dream, and he sometimes remarked to his wife: "Let us enjoy everything, before we awake."

10. *Nobel Prize, Alleged Trip to Russia*

On November 10, 1922, while Einstein was on his trip to the Orient, the committee of the Swedish Academy of Science awarded him the Nobel prize for physics. Although he had long been recognized as one of the greatest physicists of his time, it had taken rather a long while for the committee to decide to award him the prize. In establishing the endowments, Alfred Nobel had stipulated that the prize should be awarded for a recent discovery in physics from which mankind had derived a great use. No one was sure whether Einstein's theory of relativity was a "discovery." Originally, it did not assert new phenomena, but was rather a principle from which many facts could be derived more simply than formerly. Furthermore, whether this discovery was of any great use to mankind was naturally a matter of personal opinion. After Einstein's theory became an object of so many attacks and was even linked to political controversies, the Swedish Academy thought that it should be cautious and not award the prize to Einstein for a while. After the explosion of the atomic bomb in 1945, the Academy apparently recognized the great use to mankind of Einstein's theory of relativity, as it quickly awarded the prize to O. Hahn, the discoverer of the uranium fission.

Toward the end of 1922, however, the Academy thought of a clever expedient by which it could award the prize to Einstein without having to take a stand on his relativity theory. It awarded the prize to Einstein for his work in "quantum theory" (see Chapters III and IV). This work had not been so hotly debated as the theory of relativity. But in it "facts were discovered" — that is, statements were advanced from which observable phenomena could be deduced by means of few conclusions. In the case of the theory of relativity this train of reasoning was much longer. This subtle distinction, however, authorized the Academy in the case of the photoelectric and photochemical law to speak of a "discovered fact," while it would not do so in the case of the relativity theory. By this expedient the Academy succeeded to avoid the expression of any opinion about the controversial theory of relativity. The statement of the award was couched in very general terms: "The prize is awarded to Einstein for the photoelectric law and his work in the field of theoretical physics."

As soon as Einstein's enemies heard of this, they began to assert with greater vehemence than ever before that there was something peculiar about the entire business. Einstein, they said, received the prize for a discovery that was not important enough to justify such a reward. Early in 1923 his old enemy Lenard wrote a letter to the Swedish Academy in which he branded the entire action as an attempt "to restore Einstein's lost prestige without compromising the Academy itself."

In July 1923, when he received the award, Einstein lectured at a meeting of Scandinavian scientists at Göteborg, which was attended by the King of Sweden.

Since the public, especially in Germany, carefully followed everything that Einstein did, some with enthusiasm, others with suspicion and hatred, the following report, which appeared on September 15 in the *Deutsche Allgemeine Zeitung,* a paper for the more educated and wealthier nationalistic groups, could not but arouse great excitement, and in some people even anger and indignation:

"From Moscow we learn that Professor Einstein is expected there at the end of September. He will speak there on the theory of relativity. Russian scientists are looking forward to the lecture with great interest. In 1920 Einstein's writings were brought to Russia by plane, immediately translated, and appeared among the first works of the Bolshevist state press."

It must be kept in mind that in Germany Einstein's relativity theory had been characterized as "Bolshevism in physics," that many people believed in a Jewish conspiracy in which Einstein and Rathenau had participated, and finally that Rathenau had concluded the treaty of friendship with Soviet Russia. At that time the alliance with Soviet Russia was not yet regarded by the German nationalists as a particularly shrewd move in foreign policy intended to serve the national interests of Germany, but rather as a betrayal of the German people. Hence it is not surprising that many persons saw Einstein's reported trip as an indication of his participation in a Bolshevist conspiracy against Germany, and spread all kinds of rumors about it.

On October 6 the democratic *Berliner Tageblatt* reported: "Professor Einstein has left for Moscow. . . . In Moscow preparations are being made to give the famous German scientist an imposing welcome."

On October 27 the nationalistic *Berliner Börsenzeitung* reported: "The Soviet Russian press reports that Einstein is arriving in Petersburg on October 28 and will speak on the relativity theory to a group of scientifically trained workers."

On November 2 the *Kieler Zeitung* reported: "Einstein is staying in Petersburg for three days."

In the middle of November, when it was believed that Einstein had returned from Russia, he received many threatening letters in which nationalistic fanatics threatened that he would be "executed" like Rathenau, if he continued his conspiracies with the Bolsheviks. The remarkable thing about all this, however, is that Einstein has never been in Russia, either then or at any other time in his life. His journeys to France and England had frequently been taken amiss and had produced a great deal of unpleasantness for him in Germany. Evidently it was of no avail even to avoid such unpopular trips if one once had become the target of hate-filled agitators.

For Einstein the end of 1923 was the end of a period of journeys throughout the world as a messenger of international understanding and as a symbol of an omnipresent interest in the most general questions regarding the nature of the universe. In 1925 he made a trip to South America, but in general he spent the following years in Berlin.

IX

DEVELOPMENT OF ATOMIC PHYSICS

---

1. *Einstein as a Teacher in Berlin*

In 1924, after his many journeys, Einstein settled down again in Berlin. The transition from lecturing in different countries in different languages to people with various intellectual training back to regular teaching of physics was not entirely a smooth one. Since he was not required to give a regular course of lectures, he preferred to give lectures of two extremely divergent types. On the one hand, he liked to speak before an audience of educated laymen to which he could explain the general scientific principles as simply and clearly as possible, seeking to give his listeners a vivid picture of the general trends in the development of scientific thought. On the other hand, he also liked to give highly technical lectures on the problems with which he was concerned at the moment, before an audience of very advanced students.

Then, too, his world fame attracted many foreigners visiting Berlin. Their lists of sights to be seen there included, together with the Brandenburg Gate with its goddess of victory, the Siegesallee with its statues of Prussian princes, and the theatrical productions of Reinhardt, the famous Einstein. Many who did not even know whether he was a physicist, mathematician, philosopher, or dreamer came to listen to his lectures. On occasions when these sightseers were unusually numerous, Einstein would say: "Now I shall stop for a few minutes so that all those who have no further interest can leave." Usually only eight or ten students would remain, and then Einstein was happy to be able to talk about the things closest to his heart without being disturbed by the sight of faces devoid of any understanding.

Such lectures were not easy to follow even for students intending to become physicists. Even the brighter ones generally expected that Einstein would drum into their heads in a form adapted for students the famous discoveries that he had presented in his writings and about which everyone spoke. Einstein, however, was not much interested any longer in researches that had been concluded and published. He was always looking for the solutions of new problems, and students who were willing and able to think about these difficult problems independently were few and far between even in such a large center of learning as Berlin.

As I already mentioned, Einstein was at first skeptical about the use of very advanced mathematics in developing physical theories. When in 1908 Minkowski showed that Einstein's special theory of relativity could be formulated very simply in the language of four-dimensional geometry, Einstein had regarded this as the introduction of an involved formalism by which it became rather more difficult to grasp the actual physical content of the theory. When Max von Laue, in the first comprehensive book on Einstein's relativity theory, presented it in a very elegant mathematical form, Einstein remarked at that time jokingly: "I myself can hardly understand Laue's book."

The center of German mathematical teaching and research during this period was the University of Göttingen. Minkowski taught there, and the mathematical formulation of the relativity theory had begun there. Einstein once remarked playfully: "The people in Göttingen sometimes strike me, not as if they wanted to help one formulate something clearly, but instead as if they wanted only to show us physicists how much brighter they are than we." Nevertheless, the greatest mathematician in Göttingen, David Hilbert, realized that while Einstein did not care for superfluous formal difficulties in mathematics, he did know how to use mathematics where it was indicated. Hilbert once said: "Every boy in the streets of our mathematical Göttingen understands more about four-dimensional geometry than Einstein. Yet, despite that, Einstein did the work and not the mathematicians." And he once asked a gathering of mathematicians: "Do you know why Einstein said the most original and profound things about space and time that have been said in our generation? Because he had learned nothing about all the philosophy and mathematics of time and space."

In his general theory of relativity, however, Einstein had had to resort to the use of a branch of advanced mathematics called "tensor analysis" in order to give an adequate description of physical phenomena in four dimensional non-Euclidean space. With the complication in the calculations that this entailed, Einstein began to find the need for an assistant who was well trained in mathematics. For this purpose Einstein preferred young people who had a scientific education and ambition, but who because of external circumstances were unable to get a job at a public institution. Thus one of his first assistants in Berlin was a Russian Jew who suffered from a pathological enlargement of his bones (leontiasis) and as a result made such a repulsive impression on people that no one wanted to engage him as an assistant, let alone as a teacher. In time the young man understandably wanted to advance to an independent position. He expected Einstein to get him a position as teacher in a school although it was obvious that with his unfortunate appearance no school would hire him. Nevertheless he blamed Einstein for not trying hard enough and finally quarreled with him.

It was not easy for Einstein to find a suitable assistant. This may appear strange, but there were reasons for it. Students who wanted to study physics could wish for no better opportunity than to watch and help a man like Einstein at his creative work, and to this was added the pleasure of being in contact with a man with a very interesting personality, who was extremely friendly and adept in the art of conversation. But in large measure Einstein's trouble was due to the fact that he did not carry on any ordinary teaching in Berlin. The students at the university who were working toward the doctorate or to pass examinations as physics teachers were busy enough trying to satisfy all the demands made on them. They studied with the professors at the university who gave the examinations, and received from them the subjects for their doctoral dissertations. Only rarely did one of them come into personal contact with Einstein. As a result Einstein usually had as assistants students from outside Germany. These foreigners did not come to Berlin to pass examinations or to find positions, but to learn from the outstanding scientists there. They immediately turned to men like Planck, Nernst, or Einstein. In this way Einstein had as collaborators first the aforementioned Russian and later the Hungarian Cornelius Lanczos and the Austrian Walter Mayer. The last two were of great help to Einstein, and published valuable contributions to the general theory of relativity. They are now both teaching in American institutions.

2. *Structure of the Atom*

The world believed that Einstein's theory of relativity was the oddest and the most radical change in physics that had occurred for a long time. Actually new conceptions of matter even more baffling and far-reaching in their effects were being developed simultaneously.

In 1905, while still at Bern, Einstein had made outstanding contributions to the structure of light, as described in Section 10 of Chapter III. Since then he had turned his attention to his theories of relativity and gravitation, which dealt mainly with large objects such as stars and planets and not with the ultimate particle of nature — the atom. He had considered properties of light rays in gravitational fields, but in these cases it had made no difference whether light was simply a wave phenomenon or consisted of a stream of photons.

Einstein himself had realized in 1905 when he proposed the idea of light quanta (photons) that it was only a provisional hypothesis. Numerous difficulties had remained unsolved. For instance, the theory of the photon had had amazing success in explaining properties of heat radiation and the photoelectric effect, but it could not explain the whole set of phenomena dealing with the interference and diffraction of light. On the other hand, the wave theory, which could cope with these latter properties, was useless for those phenomena for which the photon theory was successful.

In conversation Einstein expressed this dual character of light as follows: "Somewhere in the continuous light waves there are certain 'peas,' the light quanta." The amplitude of the waves determines how many "peas" are present at any spot, but only as a statistical average. One can never know whether such a "pea" will be present at a particular point at a specific instant of time. From the beginning Einstein thought that this could not be the ultimate truth. "I shall never believe," he once said, "that God plays dice with the world." Nevertheless, "God's dice" penetrated into physics at several points. For instance, in the disintegration of radioactive substances a certain percentage of the atoms present disintegrate every second, but there is no way by which we can tell which particular atom will disintegrate in the next second.

But Einstein's early suggestion of "photons (light quanta) in

Five winners of the Nobel Prize in physics. This picture, taken in Berlin-Zehlendorf, shows Walter Nernst, Einstein, Planck, Millikan, and Von Laue

A recent portrait of Einstein

every light ray" had fallen on fertile soil. The "heuristic point of view" turned out to stimulate actually new discoveries. In 1913 the Danish physicist Niels Bohr attempted to correlate the *structure of atoms* with the light emitted by them. Rutherford in England had shown in 1911 that the atom consists of a central nucleus with positive charge and a number of negatively charged electrons around it. Also it had been known for a long time that free atoms, unlike glowing solid bodies that emit light with continuous distribution of different colors, emit light of only certain definite frequencies which are characteristic of the atom. In trying to explain this unique character of light emitted by free atoms Bohr found that it was completely impossible if he assumed that the electrons circulate around the nucleus according to Newton's laws of motion in the same way that the planets revolve around the sun. He was thus led to setting up a separate hypothesis, with which he modified Newton's laws in much the same way that Planck had done in explaining properties of heat radiation. Bohr assumed that only certain discrete sets of circular orbits (*preferred orbits*) were allowed for the electrons moving around the nucleus. Electrons in different orbits had different energies, and when an electron jumped from one of higher to one of lower energy, the difference in energy was emitted in the form of a light quantum (photon). This concept of emission of photons may be considered as a sort of inversion of Einstein's photoelectric law, in which a photon is absorbed and an electron liberated. But here again, as in the case of radioactive atoms, only the average behavior of the atoms, and nothing of individual cases, could be predicted. At first this deficiency did not cause much concern. It was thought that the behavior of the atoms was not unlike the mortality statistics of life-insurance companies, from which the average life expectancy of man can be predicted accurately, but not that of individuals. Nevertheless every single death has its cause. The physicists believed then that similarily causes exist for the behavior of the individual atoms, but that they are as yet unknown.

---

3. Mechanics of the Atom

The feeling that God did not play dice for the fate of the world began to be shaken about the time Einstein settled down in Berlin after his trips. In 1924 Prince Louis de Broglie, a graduate student in Paris, submitted a doctoral thesis to Professor Langevin in which he proposed even greater changes in Newtonian mechanics than Einstein had done in his theory of relativity. Langevin, who was well known as a radical in politics, was staggered by the boldness of the new proposals. De Broglie's work seemed fairly absurd to him, but considering that the idea of Bohr's "preferred" orbits was also very baffling, he thought there might be something in his student's thesis.

De Broglie had noted that Einstein's "heuristic point of view" in optics had been helpful by attributing to light properties that are usually ascribed to material particles; namely, energy and momentum of photons. De Broglie took his cue from Einstein and introduced an analogous "heuristic viewpoint" into mechanics. To resolve the difficulties in the description of the motion of *subatomic particles* (particles within the atom) de Broglie suggested that certain wave properties be attributed to particles. He assumed that, just as the motion of photons in a light ray is determined by the electromagnetic field that constitutes the light wave, so the motion of particles is guided or "steered" by a new type of waves, called "matter waves" by de Broglie and "de Broglie waves" by other physicists. According to this view, the "preferred" orbits of Bohr are orbits along which the de Broglie waves are built up by interference, while along all other orbits the waves are annihilated by interference. This phenomenon is exactly analogous to the interference patterns of light passing a small hole where there are light and dark regions depending on whether light falling on the regions from different directions builds up or interferes. De Broglie waves, however, have wave lengths inversely proportional to the momentum of the particles, and manifest themselves only in the case of very small masses, in particular in the case of subatomic particles. For any ordinary body like a billiard ball, the wave length is so small that it has no observable wave property.

Two years later Erwin Schrödinger, an Austrian, developed on the basis of de Broglie's idea a new mechanics of the atom according to which the motion of atomic particles could be calculated for any field of force. In Bohr's theory of the atom Newtonian laws and arbitrary assumptions (preferred orbits) are mixed to give satisfactory results. Schrödinger, however, obtained the same results by means of a coherent theory.

Originally de Broglie and Schrödinger had assumed that the connection between the particles and the "steering" waves by which the motion of these particles was directed was a strictly "causal" connection. But in 1926 the German physicists Max Born and P. Jordan interpreted the intensity of de Broglie waves as the average number of particles situated in a unit volume of space. The relation between the intensity of matter waves and the number of particles is thus exactly the same as that between the intensity of light and the number of Einstein's photons.

This theory, developed by de Broglie, Schrödinger, and Born, by means of which not the position itself but only the average position of atomic particles could be calculated, began to be known as *wave mechanics,* so named appropriately since it laid stress on the wave property of material particles. By this theory future observable events cannot be predicted precisely, but only statistically. For example, we cannot predict the exact point where a particle or photon will hit a screen, but only what percentage of incoming protons or particles will hit within any given region of the screen. If science could not advance beyond this stage, "God would," as Einstein said, "play dice indeed."

The idea that there are waves associated with material particles received a striking experimental verification. In 1927 two Americans, Clinton J. Davisson and L. H. Germer, proved that a beam of electrons is diffracted by a metal crystal in exactly the same way that light is diffracted by a grating, and X-rays by crystals. This confirmation is all the more amazing since diffraction is a phenomenon that is purely characteristic of waves, and nobody had ever even dreamed that it could be caused by material particles such as the electron until de Broglie suggested and Davisson and Germer actually observed it. Moreover, the wave length associated with the electrons, which could be calculated from the size of the diffraction pattern, agreed exactly with the value predicted by de Broglie.

At about the same time, W. Heisenberg, a young German, approached the interaction between subatomic particles and radiation from another direction. He broke away completely from the fundamental notion in Newtonian mechanics that a particle changes its location continuously and can thus be pursued.

Einstein in his general theory of relativity started from "Mach's requirement" that a physical theory should lead eventually to relations between quantities that can actually be measured. Accordingly, "absolute motion" was replaced by "motion relative to material bodies." Heisenberg started similarly. He abandoned the computation of the exact motions of electrons in an atom. For the laws of nature are such that it is impossible to determine the path of electrons by any measurement. The only properties of an atom that are accessible to actual measurement are the intensity and frequency of the emitted radiation. Therefore Heisenberg suggested to formulate the basic laws governing subatomic phenomena in terms of intensity and frequency of radiation. This suggestion implies a radical break with mechanistic physics, which uses "position and velocity of particles" as the basic concepts occurring in the fundamental laws of nature.

If we accept Heisenberg's suggestion, subatomic particles (like electrons or photons) are no longer "full-fledged particles" in the Newtonian sense, as their behavior cannot be described in the Newtonian way. But they are physical objects possessing some of the properties of particles.

This aproach to the theory of atoms has come to be known as *quantum mechanics.* It received a logically more satisfying form when Heisenberg went to Copenhagen and collaborated with Niels Bohr.

---

4. *Bohr's Complementarity Principle*

According to Bohr, it was not advisable to throw overboard completely the motion of particles as the basis of a description of subatomic phenomena. The presentation in terms of the observable intensity and frequency of radiation, as originally suggested by Heisenberg, should be replaced by a restricted or qualified use of the "moving particle" as the principal means of description. Heisenberg had certainly proved his point that the motion of atomic particles cannot be described in the Newtonian sense. According to Newton, once the forces that act on a particle and its initial position and momentum are given, its subsequent position and momentum at any instant can be calculated with any desired precision. Heisenberg discovered that this is not true of subatomic particles. There are no laws which connect the position and the momentum of such a particle in one instant of time with the values of these quantities at a future instant. The laws have in this domain a different character. If the initial position and momentum of a particle of very small mass (a subatomic particle) are known within a certain margin, the position at a future instant can be computed within a certain margin. By making the initial margin sufficiently narrow, however, we cannot achieve, as in Newton's mechanics, a final margin as narrow as we desire. In other words, if we want to hit a definite point of a target, we cannot be sure to achieve the desired result even if we aim very accurately. If we want to hit our target point at least within a reasonable margin, we have to consider that according to Heisenberg there is a definite relation between the initial margins of position and momentum: the product of these two margins has to equal a definite quantity, which is, roughly speaking, Planck's constant h. This relation has become famous under the name of "Heisenberg's relation of indeterminacy."

Soon afterward Bohr gave a more satisfactory interpretation of this strange behavior of atomic particles. He pointed out that "position" and "momentum" are two different aspects of a small mass (e.g., an electron) in much the same way that the particle properties and wave properties are two aspects of the photon. To say that a particle is located in a certain limited region of space is exactly analogous to the statement that light-energy is concentrated in a photon, and to define the momentum of a particle is analogous to the emphasis on the wave aspect of light. Both material particles and light have the dual characteristics of particles and waves, but their behavior is neither contradictory nor haphazard. Bohr emphasized again "Mach's requirement" that we should make only such statements as can be tested by definite physical experiments. According to him, it depends solely on the specific arrangement of apparatus used whether the emission of light and of electrons has to be described as a wave or as a beam of moving particles. According to this view, the two types of properties exhibited are "complementary" features of the same physical object. What we observe depends on what observable reaction of our subatomic phenomena we bring to a test. This conception has been called *Bohr's theory of complementarity*.

Bohr's point of view is therefore even more different from Newtonian mechanics than Einstein's theory of relativity. In Bohr's conception, we cannot describe what "actually" occurs in space while, say, light is emitted by the sun before it hits the earth. We can describe only what we observe when a measuring apparatus is hit by light. We can, for example, describe whether or not the light from the sun hits a certain spot on a screen. Or, to express it more precisely: We cannot describe "physical reality" by describing the path that a particle traverses in space, but we can and must describe only the observations made on various physical instruments arranged at different points in space and time. Physical laws link together these observations, but not the positions or paths of the particles or photons. This viewpoint was interpreted as being in agreement with *positivistic* philosophy which asserts that science cannot discover what actually happens in the world, but can only describe and combine the results of different observations.

Since the beginning of the twentieth century more and more emphasis has been placed on the conflict between the above view that science can only describe and systematize the results of observations and the view that it can and must investigate the *real world*. This controversy became particularly acute among the physicists in central Europe. Max Planck was the spokesman for the latter view, which he called the "metaphysical" view, and he directed his sharpest polemics against those who seemed to him to be the most radical representatives of the opposite side. In particular he attacked Mach's positivistic conception of science, which agrees with Bohr's view.

About this time a reformulation of positivism started in Vienna and Prague. The new movement was closely related to "Mach's requirement." The core of the movement was the *Wiener Kreis* (Vienna Circle), Moritz Schlick, R. Carnap, O. Neurath, and others. In this country it was described as *logical positivism* and established contact with related established tendencies such as *pragmatism* and *operationism*. In England a similar movement is headed by Bertrand Russell.

---

5. *Einstein's Philosophy of Science*

Since the positivistic conception of physics had been stimulated strongly by Einstein's pioneer work in the theory of relativity and in atomic physics, many persons regarded Einstein as a kind of patron saint of positivism. To the positivists he seemed to bring the blessing of science, and to their opponents he was the evil spirit. Actually his attitude to positivism and metaphysics was by no means so simple. The contradictions in his personality that we have observed in his conduct as a teacher and in his attitude to political questions also manifested themselves in his philosophy.

Einstein recognized wholeheartedly the great success of

Bohr's theory in explaining the many phenomena of atomic physics, but from a more philosophical standpoint he was not ready to admit that one must abandon the goal of describing physical reality and remain content only with the combination of observations. He was aware that it was not possible, as Newton had thought, to predict all future motions of all particles from the initial conditions and the laws of motion. But perhaps, thought Einstein, physical events could be described in terms of a new theory as yet unknown. It would consist in a system of field equations so general that they would contain the laws of motion of particles and of photons as special cases.

I must admit that over a long time I myself believed that Einstein was an adherent of the positivistic interpretation of Bohr's theory. In 1929, at a congress of German physicists in Prague, I delivered an address in which I attacked the metaphysical position of the German physicists and defended the positivistic ideas of Mach. After my address a well-known German physicist with whose philosophical views I was not acquainted rose and said: "I hold to the views of the man who for me is not only the greatest physicist of our time, but also the greatest philosopher: namely, Albert Einstein." Thereupon I felt a sense of relief and expected the speaker to support me against my opponents, but I was mistaken. The speaker declared that Einstein rejected the positivistic theories of Mach and his supporters and that he regarded physical laws as being more than combinations of observations. He added that Einstein was entirely in accord with Planck's view that physical laws describe a reality in space and time that is independent of ourselves.

At that time this presentation of Einstein's views took me very much by surprise. It was oversimplified, indeed, but I soon realized that Einstein's partly antagonistic attitude toward the positivistic position was connected with his attitude toward Bohr's conception of atomic physics. Shortly afterward I saw a paper by Lanczos, one of Einstein's closest collaborators, in which he contrasted the theory of relativity with Bohr's theory in the following manner: Einstein's general theory of relativity is the physics corresponding to the metaphysical conception of science; Bohr's theory, on the other hand, is in accord with the radical positivistic conception. I was quite astonished to find the theory of relativity characterized in this manner, since I had been accustomed to regarding it as a realization of Mach's program.

Not long afterward — I believe it was in 1932 — I was visiting

Einstein in Berlin. It had been a long time since we had conversed personally, and consequently I knew little of his stand on questions about which he had not published anything. We discussed the new physics of Bohr and his school, and Einstein said, partly as a joke, something like this: "A new fashion has now arisen in physics. By means of ingeniously formulated theoretical experiments it is proved that certain physical magnitudes cannot be measured, or, to put it more precisely, that according to accepted natural laws the investigated bodies behave in such a way as to baffle all attempts at measurement. From this the conclusion is drawn that it is completely meaningless to retain these magnitudes in the language of physics. To speak about them is pure metaphysics." In this statement, among other things, he apparently referred to magnitudes such as the "position" and "momentum" of an atomic particle.

Hearing Einstein talk in this way reminded me of many other discussions to which his theory of relativity had given rise. Repeatedly the objection had been raised: if magnitudes such as the "absolute temporal interval between two events" cannot be measured, one should not conclude that consequently it is completely meaningless to speak of this interval and that "absolute simultaneity" is simply a meaningless conglomeration of words. Einstein's reply to this argument had always been that physics can speak only about magnitudes capable of being measured by experimental methods. Furthermore, Professor P. W. Bridgman regarded Einstein's theory of simultaneity as the best illustration of the fruitfulness of his "positivistic" requirement that only magnitudes having an "operational definition" should be introduced into physics. Consequently I said to Einstein: "But the fashion you speak of was invented by you in 1905?" At first he replied humorously: "A good joke should not be repeated too often." Then in a more serious vein he explained to me that he did not see any description of a metaphysical reality in the theory of relativity, but that he did regard an electromagnetic or gravitational field as a physical reality, in the same sense that matter had formerly been considered so. The theory of relativity teaches us the connection between different descriptions of one and the same reality.

Actually Einstein has been a positivist and empiricist since he has never been willing to accept any perennial framework for physics. In the name of progress in physics he claims the right to create any system of formulations and laws that would be in agreement with new observations. For the older positivism the general laws of physics were summaries of individual observations. For Einstein the basic theoretical laws are a free creation of the imagination, the product of the activity of an inventor who is restricted in his speculation by two principles: an empirical one, that the conclusions drawn from the theory must be confirmed by experience, and a half-logical, half æsthetic principle, that the fundamental laws should be as few in number as possible and logically compatible. This conception hardly differs from that of "logical positivism," according to which the general laws are statements from which our observations can be logically derived.

In the twentieth century, when Einstein created his special theory of relativity, and even more so when he produced his general theory, it became evident that physical theories were to an ever increasing degree no longer simple summaries of observational results, and that the path between the basic principles of the theory and the observational consequences was more involved than had formerly been thought. The development of physics from the eighteenth century to Einstein was accompanied by a correspondent development of philosophy. The conception of general laws as *summaries* of observations gave way more and more to the conception that laws are creations of the imagination, which are to be *tested* by observation. *Mach's Positivism* was replaced by *Logical Positivism*.

In the Herbert Spencer Lecture which he gave at Oxford in the summer of 1933 shortly before he left Europe forever, Einstein presented the finest formulation of his views on the nature of a physical theory. He spoke first about the physics of the eighteenth and nineteenth centuries — that is, the period of mechanistic physics:

"The scientists of those times were for the most part convinced that the basic concepts and laws of physics were not in a logical sense free inventions of the human mind, but rather they were derivable by abstraction — that is, by a logical process from experiment. It was the general theory of relativity that showed in a convincing manner the incorrectness of this view."

After Einstein had emphasized that the fundamental physical concepts were products of invention or fictions, he continued:

"The conception here outlined of the purely fictitious character of the basic principles of physical theory was in the eighteenth and nineteenth centuries far from being the prevailing one. But it continues to gain more and more ground because of the ever widening gap between the basic concepts and laws on the one side and the consequences to be correlated with our experience on the other — a gap which widens progressively with the developing unification of the logical structure — that is, with the reduction of the number of the logically independent conceptual elements required for the basis of the whole system."

As in so many aspects of his life and thought, we also note a certain internal conflict in Einstein's attitude toward the positivistic conception of science. On the one hand, he felt an urge to achieve a logical clarity in physics such as had not previously been attained, an urge to carry through the consequences of an assumption with extreme radicalism, and was unwilling to accept any laws that could not be tested by observation. On the other hand, however, he felt that even *Logical Positivism* did not give sufficient credit to the role of imagination in science and did not account for the feeling that the "definitive theory" was hidden somewhere and that all one had to do was to look for it with sufficient intensity. As a result Einstein's philosophy of science often made a "metaphysical" impression on persons who are unacquainted with Einstein's positivistic requirement that the only "confirmation" of a theory is its agreement with observable facts.

6. *Unified Field Theory*

In his general theory of relativity Einstein had treated the force of gravity as due to a gravitational field. Matter gave rise to a gravitational field, which in turn acted on other material bodies to cause forces to act. Einstein had taken this force into account by means of curvature in space. A similar situation existed for electrically charged particles. Forces act between them, and they could be taken into account by considering the electric charges to give rise to an electromagnetic field, which in turn produced forces on other charged particles. Thus matter and gravitational field were exactly analogous to electric charge and electromagnetic field. Consequently Einstein sought to build a theory of "unified field" which would be a generalization of his gravitational theory and would include all electromagnetic phenomena. He also thought that in this way he might be able to obtain a more satisfactory theory of light quanta (photons) than Bohr's, and derive laws about "physical reality" instead of only laws about observational results.

The great success of the geometrical method in the general theory of relativity naturally suggested to him the idea of developing the new theory in the structure of four-dimensional space. In this case it must have still other characteristics besides the curvature which takes care of gravitational effects.

The news that Einstein was working on a unified field theory became particularly widespread in 1929, the year of Einstein's fiftieth birthday. To the public at large it seemed to be an especially attractive idea that on the very day on which he attained fifty years, a man should also find the magic formula by which all the puzzles of nature would finally be solved. Einstein received telegrams from newspapers and publishers in all parts of the world requesting that he acquaint them in a few words with the contents of his new theory. Hundreds of reporters beseiged his house. When some reporters were finally able to get hold of him, Einstein said with astonishment: "I really don't need any publicity." But everyone expected some new sensation that would surpass the wonder produced by his previous theories. They learned that a communication dealing with the new theory would be published in the transactions of the Prussian Academy of Science, and efforts were made by newspapers to secure galley proofs from the printer, but without success. There was nothing to do but to await the publication of the article, and in order not to be too late, an American newspaper arranged to have it sent immediately by phototelegraphy.

The article was only a few pages long, but it consisted for the most part of mathematical formulæ that were completely unintelligible to the public. The emotion with which it was received by the layman may be compared to that experienced at the sight of an Assyrian cuneiform inscription. For an understanding of the paper a considerable capacity for abstract geometrical thinking was required. To those who possessed this quality it revealed that general laws for a unified field could be derived from a certain hypothesis regarding the structure of four-dimensional space. It could also be shown that these laws included the known laws of the electromagnetic field as well as Einstein's law of gravitation as special cases. Nevertheless, as yet no result capable of experimental verification could be derived from them. Thus for the public at large the new theory was even more incomprehensible than the previous theories. For the expert it was an accomplishment of great logical and æsthetic perfection.

X

POLITICAL TURMOIL IN GERMANY

1. *Einstein's Fiftieth Birthday*

As the month of March in 1929 approached, Einstein and his family began to fear that the sensationalism of the newspapers would be so great on the date of his fiftieth birthday that it would only be disagreeable for Einstein. Many newspapers had undertaken to secure Einstein's own remarks on more or less personal matters and to publish them. Moreover, the visits and congratulations of his true admirers and friends threatened to assume such proportions that Einstein decided to avoid everything and to leave his apartment for several days. Immediately all sorts of rumors appeared: Einstein has gone to France, to Holland, to England, or even to America. But it was all greatly exaggerated. He spent the day peacefully near Berlin at the country estate of a shoe-polish manufacturer, who sometimes put at Einstein's disposal a pavilion in his garden, situated very close to a beautiful lake. Here he was able to play the organ or to sail on the lake.

From their apartment in Berlin Mrs. Einstein had brought along the dinner that had been prepared. Einstein's immediate family — that is, his wife, her two daughters, and their husbands — were present. Einstein was very comfortable and unceremonious, dressed in the garb he usually wore in the country, or even in the city when no strangers were present. This consisted of a pair of old trousers and a sweater, but no jacket, and very often also without shoes or stockings. From their city apartment Mrs. Einstein also brought along some of the congratulatory letters and presents that had arrived in large numbers.

Einstein was connected with many different activities, so that he received letters and gifts from all sorts of people; naturally, from physicists and philosophers, but also from pacifists and Zionists. There were even some from very simple people who were admirers of great discoveries and wanted to express this admiration. Among these was a gift from an unemployed man, consisting of a small package of pipe tobacco. It had become

220

generally known that Einstein was rarely to be found without a pipe. Alluding to the relativity theory and the field theory, the man wrote: "There is relatively little tobacco, but it is from a good field."

Several of his friends had combined to present him with a new and very modern sailboat. Einstein loved to sail the beautiful lakes and rivers around Berlin, and to daydream while the boat flew before the wind. The handling of the sails was a pleasant activity. It was a very simple application of the rules of mechanics, and it gave him a great deal of pleasure to apply the physical laws that are closest to direct experience instead of those that are most abstract. He also wrote a popular article in which he explained to the lay public the physical laws that enable one to travel in a certain direction by placing the sails in a certain position and to reach a particular goal by means of a zigzag motion — that is, by successive tacks.

A group of Zionists in America bought a plot of land in Palestine and planted it with trees on his birthday. They made provision that for all time to come the woods that grew there were to be known as the Einstein Grove.

The most beautiful and interesting present, however, was to come from the municipal administration of the city of Berlin, where Einstein had lived since 1913, and which, to mention only a very trivial matter, he had helped to make a center of attraction for all foreigners. Since it was generally known that Einstein was fond of sailing on the Havel River and on the many lakes into which this remarkable stream expands, the municipal council of Berlin decided to present Einstein with a small country house situated on the bank of the Havel close to the point where it enters the Wannsee. The house was located on a plot belonging to the city of Berlin. This resolution on the part of the municipal council was well received by the entire population — a sentiment arising from a combination of love of science, respect for an illustrious fellow citizen, and a fondness for aquatic sports and sailing. In all the illustrated magazines appeared pictures of the idyllic "Einstein house."

When Mrs. Einstein wanted to see the house, she noticed to her amazement that people were living in it. The latter, in turn, were astonished to find someone wanting to take possession of their home, even though it was the famous Einstein. It turned out that when the city of Berlin had acquired this property, it had guaranteed to the inhabitants of the house the right to keep on living there. The municipal council seemed to have forgotten this when it gave its birthday present to Einstein. How can one explain such an occurrence in Berlin, the capital of Prussia, famous for its orderliness?

At first it seemed to indicate a considerable confusion in the registry of landed property. When the leaders of the municipal council heard about this mistake, they wanted to correct it as soon as possible. The park in which the frustrated "Einstein house" stood was large and filled with beautiful trees, and there was enough room in it for several houses. The council therefore chose another part of the park, very close to the water, and offered it to Einstein as a birthday present. The house, however, was to be built at his own expense. Einstein and his wife were very happy about it and agreed to this arrangement. But on closer investigation it was found that this was also impossible. When the owner of the "Einstein house" received the right to live in it, he had also been assured that no other house would be built in the park that might in any way disturb his enjoyment of nature and his view over the lake.

Finally the entire matter began to become unpleasant for both Einstein and the municipal council. A gift that came into being in this way could no longer give pleasure to anyone. Thus it became more and more of a mystery what was actually occurring in the famous model city of Berlin.

But the matter was not yet at an end. After considerable reflection the municipal council hit upon a third piece of land near the water. It was not nearly so well situated nor was it actually near the water. The neighbors, however, permitted at least a passage from the piece of land in question to the water. The gift became poorer and poorer. When it was finally discovered that the city had no right to dispose of this third plot of land, all Berlin burst out laughing. The laughter aimed at the municipal administration was justified, but Einstein was involved in the matter through no fault of his own.

Now the council finally became aware that there was no land whatever at its disposal along the water. But since the magnificent gesture of presenting a gift to the Berlin scientist had already become public knowledge, the members of the council felt ashamed to let the entire matter turn into a fiasco. A delegate came to Einstein and said: "In order to be sure that the land we will present to you really belongs to us, please pick out a plot of land that suits you and is for sale. We will buy it." Einstein agreed. But since he did not like to occupy himself with choosing a piece of land, he let his wife go out to look. Finally she found a beautiful place in the village of Caputh, near Potsdam. The council agreed to the selection, and at the next session of the council a motion for the purchase of the land was presented. Thereupon the entire matter began to develop into a political dispute. A representative of the nationalist parties began to discuss whether Einstein actually deserved such a gift. The subject was postponed to the next session.

Then Einstein finally lost patience. The gift from his adopted city presented in the name of all the citizens had become an object of political strife, and under the most favorable circumstances it would result from a political bargain. Einstein wrote a letter to the Mayor of Berlin, who later occupied a prominent place in the public eye when it became known that he had accepted a gift of a fur coat for his wife from persons to whom he had given municipal contracts. Einstein wrote approximately as follows: "My dear Mr. Mayor: Human life is very short, while the authorities work very slowly. I feel therefore that my life is too short for me to adapt myself to your methods. I thank you for your friendly intentions. Now, however, my birthday is already past and I decline the gift."

The result of the entire matter was that Einstein not only built the house at his own expense, but also had to buy the land with his own money. Some time after these events I was in Berlin and Mrs. Einstein said to me: "In this way, without wanting it, we have acquired a beautiful home of our own situated in the woods near the water. But we have also spent most of our savings. Now we have no money, but we have our land and property. This gives one a much greater sense of security."

This feeling was to be proved wrong, because hardly three years later Einstein and his wife had to leave the land and their beautiful villa with its new furnishings. This, however, is more a private matter. Much more interesting is the question of how this entire comedy of errors was possible in the orderly city of Berlin. The answer to this question is the answer to the whole problem of the German Republic. The city of Berlin was apparently headed by men who represented culture and who wished to express this position by honoring Einstein. The decisive power, however, lay in the hands of persons who sabotaged the work of the apparent rulers. The officials of the city of Berlin carried out the orders of the municipal council in such a way as to result in failure and to make the republican administration look ridiculous.

The situation was similar throughout the German Republic.

The Chancellor and the government showed their admiration for art and science; but even at that time the real power already lay in the hands of the underworld.

---

2. *Visiting Professor at Pasadena*

In the following year, 1930, Einstein received an invitation to spend the winter in Pasadena, California, as visiting professor at the California Institute of Technology. Consequently in December he sailed for America. At this time his entire political interest was concentrated on pacifism, and he felt that this was also the great mission of the United States. While still on shipboard he broadcasted a message to America in which he said:

"Greetings to America. This morning, after an absence of ten years, when I am once more about to set foot on the soil of the United States, the thought uppermost in my mind is this: This country has through hard labor achieved the position of undisputed pre-eminence among the nations of the world. . . . It is in your country, my friends, that those latent forces which eventually will kill any serious monster of professional militarism will be able to make themselves felt more clearly and definitely. Your political and economic condition today is such that you will be able to destroy entirely the dreadful tradition of military violence. . . . It is along these lines of endeavor that your mission lies at the present moment. . . ."

Einstein was not of the opinion, however, that the United States could accomplish this mission by a policy of isolation. On March 29, 1931 he wrote: "In this country the conviction must grow that her citizens bear a great responsibility in the field of international politics. The role of passive spectator is not worthy of this country." Moreover, he always regarded America's intervention in world politics as an intervention in favor of peace. He quoted Benjamin Franklin, who had said: "There never was a bad peace or a good war."

This time Einstein did not have to make such troublesome and disturbing trips throughout the entire country. Instead he was invited to take part in the scientific research that was being carried on at the California Institute of Technology and the Mount Wilson Observatory. Both institutions are situated near Pasadena, a quiet suburb of Los Angeles. Through the efforts of R. A. Millikan, the California Institute of Technology had become a center of physical research. Millikan, a recipient of the Nobel prize, was originally a student of Michelson, and was consequently acquainted with the entire trend of Einstein's research from its experimental aspect. He has been a man possessing not only scientific, but also administrative ability, and he has always been a realist. Einstein's enthusiasm for pacifism always appeared to him as something not suited to our world, and this opinion was to be proved correct only too soon. Millikan was in accord with Einstein on one point, however: neither of them denied the important role of religious communities in the advancement of human co-operation. But neither Millikan nor Einstein recognized any control over science by religious dogmas.

In the spring of 1931 Einstein returned to Berlin and in the fall went back again to Pasadena to spend another winter there. When he again returned to Berlin in the spring of 1932, he arrived just in time to witness the principal act in the death agony of the German Republic.

In March 1932 a presidential election was to take place. The Imperial field marshal, the octogenarian Hindenburg, was the candidate of the Democrats and Socialists; his chief opponent was Adolf Hitler, the leader of the Right-radical revolutionaries. Thanks to the propaganda of Reich Chancellor Brüning, Hindenburg won the election. The Republicans and Democrats were jubilant, but the truth was that now the power was in the hands of an adherent of the former German monarchy. Under the influence of his immediate environment, he used the power to overthrow the Republic.

Hindenburg's first act after his election in May was to compel Brüning, his most faithful champion and the man who had brought about his election, to resign as Chancellor. In his place he appointed Papen, a man who was resolved to rule with the support of bayonets and to eradicate every trace of republicanism and democracy. He announced to the Reichstag that a "fundamentally new regime" was beginning, now that the period of "materialism" was at an end. With the aid of the Reichswehr, he deposed the Prussian government.

Many scientists were happy at these developments. They believed that now the reins were in the hands of the military. Since the time of Bismarck they had been accustomed to the belief that for Germany as a state and people, the rule of the "professors" could only be harmful. The fall of the "intellectuals and democrats" would enable Germany to become great.

I can still recall very well a conversation that I had with Einstein in the summer of 1932. We were at his country home in Caputh. It was a log house, constructed of sturdy beams, and we looked out through enormous windows on the idyllic forest landscape. When a professor who was present expressed the hope that a military regime might curb the Nazis, Einstein remarked: "I am convinced that a military regime will not prevent the imminent National Socialist revolution. The military dictatorship will suppress the popular will and the people will seek protection against the rule of the junkers and the officers in a Right-radical revolution."

Someone asked Einstein for his opinion of Schleicher, the "social general" who would perhaps soon seize power. "He will produce the same result as the present military dictatorship," Einstein replied.

During this summer Abraham Flexner, the famous American educator, came to Caputh to interest Einstein in his new research institute at Princeton. "For the time being," said Einstein, "I am still under obligation to spend the coming winter in Pasadena. Later, however, I shall be ready to work with you."

When Einstein set out with his wife for California in the fall of 1932, and as they left the beautiful villa in idyllic Caputh, Einstein said to her: "Before you leave our villa this time, take a good look at it."

"Why?" she asked.

"You will never see it again," Einstein replied quietly. His wife thought he was being rather foolish.

In December, Schleicher became Chancellor. He wanted to form a new government based on the working class, but the power of President Hindenburg was exerted against him. Schleicher was only a transitional phase. At the end of January 1933, while Einstein was still in sunny California discussing with the astronomers of Mount Wilson Observatory the distribution of matter in space and similar problems of the universe, Schleicher resigned and President Hindenburg appointed Adolf Hitler, his opponent at the last presidential election, as Chancellor of the German Reich.

3. *Racial Purging in German Universities*

Heretofore no aspect of Marxism had been so repugnant to the German professors as the assertion that the evolution of scientific knowledge is influenced by political power. Their highest ideal was always the complete independence of science from politics and the sharp separation of the two. But now the political power had come into the hands of Chancellor Hitler and his party whose foremost principle was the primacy of politics over all fields of human life; over science just as much as over economic life, art, and religion.

The standpoint of the new government is understandable if one remembers that the new state not only appeared as a new political organization, but also claimed to represent a new philosophy and a new orientation in all fields of life. The new orientation was that every effort was to be directed toward the goal of serving the German people and the German race. This was the ultimate aim of science just as it was of any other activity.

This conviction that an entirely new *Weltanschauung* had to be taught at the universities led the government to put pressure on the university teachers. But since the freedom of science was one of the most favored slogans in the professorial world, the new government sought to introduce its goal by compulsion while retaining the old mode of expression as far as possible. The fine-sounding word "freedom" continued to be used, but it received a new meaning. The equivocal use of this word in earlier German philosoph y had already prepared the ground for the National Socialist use. In an essay on "German Freedom" written under the influence of the first World War, the American philosopher George Santayana had already said:

"Freedom in the mouth of German philosophy has a very special meaning. It does not refer to any possibility of choice nor any private initiative. German freedom is like the freedom of the angels in heaven who see the face of God and cannot sin. It lies in such a deep understanding of what is actually established that you would not have it otherwise; you appropriate and bless it all and feel it to be the providential expression of your own spirit. You are merged by sympathy with your work, your country and the universe, until you are no longer conscious of the least distinction between the Creator, the state and yourself. Your compulsory service then becomes perfect freedom."

A clear presentation of the practical application of this profound metaphysical theory was given by E. Krieck, German pedagogical leader at this period:

"It is not science that must be restricted, but rather the scientific investigators and teachers; only scientifically talented men who have pledged their entire personality to the nation, to the racial conception of the world, and to the German mission will teach and carry on research at the German universities."

Thus a philosophical foundation was provided for the "cleansing" of the faculties of the German universities.

The first application of the new theories was in the eradication of all teachers at institutions of higher learning who on the basis of their racial origin were not considered fit to train the youth in the spirit of the new philosophy. In this group were all those who did not belong to the Germanic or Nordic, or, as it was frequently called, the Aryan race. This grouping of non-German or non-Aryans was meant specifically for the Jews, since it was believed that because of their history and education they formed a group that would tend to hinder the training in the spirit of the new rulers. The term "Jews" included not only those who professed the Jewish religion. The new government assumed a standpoint of neutrality toward religion as such. What the National Socialists meant was the Jews as a race; but in this case there was no clear criterion by which to determine a racial Jew. Since such a definition was difficult to make and had to be arbitrary in some degree, the conscientious and thorough German professors believed that no racial "cleansing" could take place. Without a neat and tidy definition the German government would be unable to do anything.

But they were still unacquainted with the "pragmatic" spirit of the new philosophy. The definitions that were needed were produced with the greatest speed, even though they did not satisfy the requirements of the German professors with respect to anthropological, ethnological, or philological accuracy, or even logical consistency. From the very beginning it was obvious that there was no scientific definition of an "Aryan," except that he was a person who spoke a language belonging to the "Aryan linguistic family." Such a definition, however, was impossible; otherwise everyone who spoke Yiddish, which is basically a German dialect, would be an Aryan. Thus from the beginning it was not the "Aryan," but rather the "non-Aryan" who was defined. The definition of a non-Aryan included everyone who had at least one non-Aryan grandparent. The grandparents, however, were defined as non-Aryan if they professed the Jewish religion; they were defined, that is, in terms of a criterion that has nothing to do with race in the ethnological sense. It was simply taken for granted that two generations earlier there were no persons of Jewish origin who professed the Christian religion.

This cunning combination of definitions on the basis of origin and religion achieved the intended political purpose: namely, to exclude an entire group of people that it was feared could exert a dangerous political or ideological influence on the students. The definition, however, was not characterized by the scientific clarity and precision that the professors required. Quite a few would have been ready to co-operate in carrying out a political purge of the universities, but it would have to be done in a scientifically unobjectionable manner.

The attempt to exclude the Jews everywhere, but to talk only of non-Aryans, gave rise to many difficulties. According to the customary meaning and usage of the word "Aryan" prior to the advent of the Nazis, there were other non-Aryans besides the Jews. At first rather unpleasant sensations were aroused by the idea that such peoples as Hungarians and Finns, who were very popular with the National Socialists, were to be branded as non-Aryans. On the other hand, one could not very well call a Hungarian an Aryan. Consequently, it was decided that a non-Aryan status is determined by means of the official definition using the religion of the grandparents. Nevertheless, even if anyone — a Hungarian, for instance — can prove that he is not a non-Aryan, it does not follow that he is an Aryan. Thus one of the fundamental rules of ordinary logic was dropped: namely, the principle of the excluded middle, which says that a thing either has or does not have a certain characteristic; there is no other possibility. According to the new official mode of expression, however, a Hungarian was neither a non-Aryan nor an Aryan.

As the new regime achieved political successes the number of people who were neither Aryans nor non-Aryans grew ever greater. The Japanese were soon the outstanding members of this group. Finally, however, when their anti-British policy led the National Socialists to seek the friendship of the "Semitic" Arabs, the latter were also included among the "non-non-Aryans." Previously the Jews had been opposed because, it was said, they belonged to the "Semitic" race. Now, however, with the inclusion of this race among the noble races, it was asserted that the Jews did not belong to any race at all, but instead formed a mongrel "anti-race."

But since a criterion of race that was not based on a religious confession was still wanted, it was finally decided to consider as related to the German race every other race that lived in "compact settlements" and not, like the Jews, scattered in separate cities and commercial centers.

The definition that had been so anxiously awaited was thus successful, and the universities were thoroughly purged according to this pattern. At first there were still several exceptions. All those professors were retained who had been appointed by the Imperial German government and not by the Republic, because it was assumed that the latter had favored the Jews. Furthermore, all those were allowed to remain who had fought for Germany or her allies in the World War of 1914–18.

In time, however, all these exceptions were dropped and the purge became more rigorous. Soon a further step was taken, and all teachers were dismissed whose wives were non-Aryans according to the official definition.

The "racial" purge was accompanied by a simultaneous political purge. But the principles upon which it was based were much less distinct. The professors who were dismissed included all those who had taken an active part in the work of the Social Democratic and Communist parties, or who had belonged to the Freemasons or to a pacifist organization. All other principles were vague. This purge was even more baffling than the racial one, since in the latter case the individual's fate was predestined and he could do little to improve it. On the other hand, by means of good behavior anyone could hope to make good any previous political sins. Thus many professors who were formerly known as "democrats" now began to express in a very obtrusive manner their sympathy with the racial purge and with other catchwords of the ruling party. Or one saw such democratic sinners engaged in studying the application of the race theory to such fields as mathematics, chemistry, and so forth. On the other hand, many former supporters of the old nationalist and monarchist groups assumed an attitude of reserve toward the new masters. Actually some of those who had been victims of the first political purge were later reinstated after they had shown signs of "improvement."

In order to make the change even more thorough, advantage

was taken of this opportunity to pension off, because of age, many older professors who were not suspect on either racial or political grounds. It was believed that they would be unable to adapt themselves to the new regime. As a result of all these measures it was possible to appoint many new teachers whom the government considered reliable and who it was believed would teach in the light of the new philosophy.

---

4. *Hostility toward Einstein*

When the purge began, Einstein was fortunately not in Germany. It was immediately evident, however, that the hostility of the new rulers to certain scientific groups was concentrated to an astonishing and even frightening degree against Einstein. Just as the general enthusiasm for Einstein's theory is an amazing phenomenon in the history of science, so the persecution of a man who advanced such abstract theories is likewise very puzzling.

His opponents may have said: "He is a Jew and became world-renowned as a creator of new ideas. This is not in accord with the views of the new rulers on the intellectual sterility of the Jewish race. He is a pacifist and sympathizes with the efforts for international co-operation." Nevertheless, this does not suffice to explain the intensity of the antagonism to Einstein. Here as well as in the growth of his fame a process of crystallization was involved. Hate is added to hate, and fame to fame, just as new crystals arise by forming around already existing crystals.

This development finally reached a point where the National Socialists believed that Einstein was the chief of a secret movement, sometimes described as "communistic," sometimes as the "Jewish International," which was working against the new government.

Actually Einstein has always steered clear of actual politics. The National Socialists, however, not only set about to attack Einstein's purely theoretical remarks on politics, which were in general academic, but they also tried to show that there was something "Bolshevistic" and "Jewish" about his theories.

As we have seen, the modest beginnings of these attacks were already evident at the end of the war in 1918. Now, however, the leaders of the campaign against Einstein felt that their time had come. Now they could come out into the open with their

sincere opinions, while Einstein's defenders were no longer allowed to reply to them. Thus in May 1933 Lenard, Einstein's
old enemy, published an article in the *Völkische Beobachter,* the
chief organ of the National Socialist Party. Here Lenard could
finally speak without having to restrain himself in any way:

"The most important example of the dangerous influence of Jewish circles on the study of nature has been provided by Herr Einstein
with his mathematically botched-up theories consisting of some ancient knowledge and a few arbitrary additions. This theory now
gradually falls to pieces, as is the fate of all products that are estranged
from nature. Even scientists who have otherwise done solid work
cannot escape the reproach that they allowed the relativity theory to
get a foothold in Germany, because they did not see, or did not want
to see, how wrong it is, outside the field of science also, to regard this
Jew as a good German."

Two years later this same Lenard delivered an inaugural address at the opening of a new physics institute in which he said:

"I hope that the institute may stand as a battle flag against the
Asiatic spirit in science. Our Führer has eliminated this same spirit
in politics and national economy, where it is known as Marxism. In
natural science, however, with the overemphasis on Einstein, it still
holds sway. We must recognize that it is unworthy of a German to
be the intellectual follower of a Jew. Natural science, properly so
called, is of completely Aryan origin, and Germans must today also
find their own way out into the unknown. *Heil Hitler.*"

Proof that Einstein's research was characteristically "Jewish"
was obtained by producing a definition of "Jewish physics" that
contained all the characteristic features of Einsteinian physics.
Thus it was regarded as particularly "Jewish," if a theory was
very "abstract"; that is, if it was connected with the immediate sensory observations only by long trains of thought and
did not lead to immediate technical applications. All this was
now considered "Jewish." It had been completely forgotten that
innumerable adherents of the Nordic doctrine had proved that
the Aryan spirit hovers in the heaven of speculation, while the
"non-Aryan" is at home in the material world which is the only
one that he comprehends with his "inferior mind."

The demand that science occupy itself with immediate practical necessities is not uncommon in a new regime that must
develop the resources of a country as rapidly as possible, whether
it be for a policy of conquest or reconstruction. We find similar
features at the beginning of the Soviet regime in Russia.

In 1934 Hermann Göring, the second leading Nazi, said:

"We honor and respect science; but it must not become an end in itself and degenerate into intellectual arrogance. Right now our scientists have a fertile field. They should find out how this or that raw material that we must import from abroad can be replaced equally well at home."

And the Minister of Education Bernhard Rust said briefly and succinctly: "National Socialism is not an enemy of science, but only of theories."

Herewith not only Einstein himself, but actually an entire science, theoretical physics, was condemned. At about this time an outstanding representative of this science in Germany who had been spared in the purge remarked to me jokingly: "You must know that Einstein has compromised our entire science."

Only a few years previously the German physicist Wilhelm Wien, who was rather sympathetic toward German nationalism, in a conversation with the great English physicist Ernest Rutherford, had said: "The relativity theory is something that you Anglo-Saxons will never understand, because it requires a genuine German feeling for abstract speculation." And the nationalistic French physicist Bouasse said: "The French spirit with its desire for Latin lucidity will never understand the theory of relativity. It is a product of the Teutonic tendency to mystical speculation."

As I have said, when the great purge began, Einstein was still in America. Upon learning of the events in Germany he went to New York and communicated with the German consul. In accordance with his official duty the latter told Einstein that he need have no fear to return to Germany. A "national" government was now in power there, which would do justice to all. If he was innocent, nothing would happen to him. Einstein had decided, however, not to return to Germany so long as the existing regime remained in power; and he said so quite openly at the consulate. After the official conversation was at an end, the consul's deputy said to him privately: "Herr Professor, now that we are speaking as man to man, I can only tell you that you are doing the right thing."

Numerous reporters wanted to hear Einstein's opinion on the recent events in Germany. But he repeated what he had always said: he had no desire to live in a state where freedom of expression did not exist and in which racial and religious intolerance prevailed. He did not enter into any concrete discussions, however.

He sailed for Europe, and in the spring of 1933 took up his

residence in the Belgian sea resort Le Cocque, not far from Ostend. From the beginning he knew that his connection with the Prussian Academy must cease. The only question was whether he should resign of his own accord or wait until the Academy expelled him. The leading person in the Academy at this time was Max Planck, the man who had first "discovered" Einstein, who had declared him to be the Copernicus of the twentieth century, and who, despite all conflicts, had supported him all the time that he was in Berlin. One can imagine that this man did not want to exclude Einstein from the Academy. And in turn, Einstein wanted to spare him this unpleasant step. He wrote briefly and to the point that under the present government he could no longer serve the Prussian state and therefore resigned his position.

At first the Academy beat about the bush, and there were great discussions about what should be done. On one hand was the desire to retain the reputation of the Academy as an impartial scientific body, on the other the desire to avow the idea of the national government. Nernst, who was always something of a liberal, said at one session: "Why should one demand of a member of the Academy, who is a great mathematician, that he should also be a nationally minded German? Were not d'Alembert, Maupertuis, and Voltaire members of our Academy, of whom we are proud even today? And these men, moreover, were Frenchmen." He repeated over and over again, when he met an academician: "How will posterity judge our Academy? Won't we be regarded as cowards who yielded to force?"

But since the newspapers of the ruling party were already full of attacks against Einstein and accused him of agitating against his own country abroad, the Academy finally decided to publish a statement, characterized by a certain dolefulness, in which they denied having any connection with Einstein. "We have no reason to regret Einstein's resignation," it said. "The Academy is aghast at his agitational activities abroad. Its members have always felt in themselves a profound loyalty to the Prussian state. Even though they have kept apart from all party politics, yet they have always emphasized their loyalty to the national idea."

Einstein who was unaware that he had been actively engaged in agitation abroad, answered the Academy in a letter on April 5:

"I am not aware that I have spread so-called 'atrocity stories' about Germany abroad. And, to be honest, I have not ever noticed that any 'atrocity reports' were being circulated. What I have noticed is that the statements made by members of the new German government have been repeated and commented upon, especially the program for the destruction of the German Jews. . . . I hope that the Academy will transmit this letter to its members and will also do its part to spread it among the German public; because I have been libeled in the press, and the Academy by its communications to the newspapers has assisted this libel."

Since the Academy could no longer assert that Einstein had invented "atrocity stories" and spread them abroad, it retreated to the statement that while Einstein had not invented any stories, yet he had made no effort to oppose energetically those that were in circulation and to defend his fatherland.

On April 7 the Academy wrote to Einstein approximately as follows:

"We have awaited with confidence for a man like you, who was for so long a member of our Academy, to range himself at the side of our nation and without regard for his political sympathies to oppose the flood of lies that has been let loose against us. In these days when filth is hurled at the German nation, partly in a common, partly in a ridiculous manner, a kind word for Germany from the mouth of such a famous man as you would have had a great effect abroad.

"Instead, your remarks were still another instrument for the enemies not only of the present German government, but also of the entire German people. This was a bitter disappointment to us. It would have led to a parting of our ways under any circumstances, even if we had not received your resignation."

Einstein now saw that a continuation of the correspondence would have no further purpose. On April 12 he wrote a farewell letter to the Academy, with which he was linked by so much work in common. In it he said:

"You write that a kind word on my part concerning the German people would have had a great effect abroad. To this I must reply that such a 'kind word' would have been a denial of every concept of justice and freedom for which I have fought all my life. Such testimony would not have been, as you put it, a 'kind word' for the German people. On the contrary, such a statement would only have helped to undermine the ideas and principles by means of which the German people have acquired an honorable place in the civilized world. By such a testimony I would have contributed, even though indirectly, to the barbarization of morals and the destruction of cultural values.

"Your letter shows me only how right I was to resign my position at the Academy."

Einstein had voluntarily resigned from his position at the Academy in order to spare a man like Max Planck the painful and shameful act of expelling from the Academy at the behest of a political party a man whom he himself regarded as one of its most valuable members. Max Planck was one of the German professors who repeatedly asserted that the new rulers were pursuing a great and noble aim. We scientists, who do not understand politics, ought not to make any difficulties for them. It is our task to see to it that as far as is possible individual scientists suffer as few hardships as possible, and above all we should do everything in our power to maintain the high level of science in Germany. At least envious foreigners should not notice that a lowering of the level is taking place anywhere in our country.

The idea that the many brutalities practiced upon individuals and institutions were only temporary attendant phenomena of the "revolution from the Right" was widespread among men of Planck's type. One of the outstanding scientists of the University of Berlin approached Planck and told him that he would like to leave Berlin immediately and look about for a place to work abroad. He felt that one day he would become a victim of a later purge. To this Planck replied: "But, my dear colleague, what strange ideas you have! If you do not find present conditions at the universities congenial, why don't you take a leave of absence for a year? Take a pleasant trip abroad and carry on some studies. And when you return all the unpleasant features of our present government will have disappeared."

At the Kaiser Wilhelm Institute, of which he was president, Planck likewise endeavored to retain non-Aryan scientists in their positions. He believed that in this way he would be able to prevent those people whom he valued from suffering hardships. As a result the research work of the institute and the fame of German science would, he hoped, remain undamaged.

He was helped by the circumstance that non-Aryans were tolerated somewhat longer in the field of research than in the teaching profession. Thus Planck succeeded in retaining several of these research workers even after the general purge in Berlin. But when the purge finally did hit them, they were in a particularly poor situation. It was more difficult for them to leave Germany and to find positions abroad.

Planck once tried by personal intervention to convince Adolf Hitler that the mechanical application of his "non-Aryan definition" to the organization of education and research would have an unfavorable effect. Planck's meeting with Hitler was the subject of much discussion in Berlin university circles at this time. Planck had but little opportunity to present his arguments. The Führer spoke to him in an argumentative manner as if he were spreading agitational propaganda at a mass meeting, and not as one speaks to a single visitor in an office. Among other things, Hitler said that he would give the Jews some opportunity to work if they were not all Bolsheviks. When Planck timidly objected that this certainly did not apply to a man like Haber, Hitler said: "Believe me. Those that are not Bolsheviks openly are so in secret." Furthermore said the Führer decisively: "Do not think that I have such weak nerves as to let myself be diverted from my great goal by such petty considerations. Everything will be carried out to the last letter."

As we have seen, Einstein had spared the Prussian Academy the embarrassment of having to throw him out, but he did receive an official letter from the Bavarian Academy of Science informing him that he was expelled from its ranks.

Einstein's villa at Caputh was searched by the political police. It was believed that the Communist Party had hidden stores of arms there. Such accusations were a result of the fantastic ideas regarding Einstein's role as a political leader or conspirator.

Einstein's possessions, his villa as well as his bank account, were all confiscated by the state. In the announcement of this act that he received from the political police the reason given was: "The property was *obviously* going to be used to finance a Communist revolt." The "gift" of the city of Berlin had led him to use the greatest part of his savings to build his villa, which was now confiscated, and Einstein had but little left of all his property. Simultaneously it became evident that by adopting German citizenship as a sign of sympathy for the German Republic he had acted to his own disadvantage, since as a foreigner (Swiss) he would have been protected against the confiscation of his property.

Einstein's writings on the relativity theory were burned publicly in the square before the State Opera House in Berlin, together with other books, some of which were regarded as obscene, others as Bolshevistic. For some time there was even a regulation according to which all books written by Jews were to be marked "translated from the Hebrew." This was intended to express that they were only apparently written in German. At that time there was still an occasional professor of physics in

Germany who while lecturing on the relativity theory permitted himself the joking remark: "It is a mistake to believe that Einstein's original paper was translated from the Hebrew."

As was to be expected, some of Einstein's scientific opponents took advantage of the new regime's hostility toward him to prevent as far as possible the teaching of Einstein's theories at the German universities. Among these opponents, in addition to the aforementioned Lenard, was another well-known physicist named Johannes Stark. He had made some outstanding experimental discoveries, for which, like Lenard, he had received the Nobel prize. But he was just as incapable as Lenard of comprehending a complex theoretical structure. Like Lenard he advocated the view that there was something "un-German" in the predominance of theory over sensory observation and it must therefore be eradicated from the teaching in the German schools. Stark also found an explanation for the fact that so many German physicists accepted the relativity theory even though it was repugnant to the German spirit. He explained it as resulting from the circumstance that so many physicists had Jewish wives.

This use of political power to compel the acceptance of one view in the field of science aroused great concern among the German physicists. One of the leading physicists said to me at that time: "It is fortunate for us that Lenard and Stark are no longer young. If they still had their youthful *élan* they would command what should be taught as physics."

Nevertheless, not everything was carried out as radically as Einstein's opponents wanted it. The National Socialist Party even adopted a resolution stating that no physical theory could claim to be "genuinely National Socialistic." Thus Einstein's theory was not completely eradicated in the German universities. It depended on the courage of the individual teacher. Some taught the theory without mentioning Einstein's name, others dropped the name "relativity theory." Others went still further; they taught the individual facts that followed from this theory as facts of experience, but they omitted completely the logical connection of these points by the theory. No physicist could dispense with these important facts, such as the relation between mass and energy or between mass and velocity.

Most of the German physicists were at their wits' end thinking up ways of protecting themselves from the continual interference in their science by the political physicists such as Lenard, and some of them hit upon an idea that despite the seriousness of the situation had something comical about it. They thought that there was only one way to shake Lenard's prestige with the new authorities, and that was to prove that he was a non-Aryan. This seemed a plausible possibility, as Lenard's father had conducted a brokerage business in Pressburg (Bratislava), the capital of modern Slovakia. Since many of the inhabitants of this city were Jewish and the brokerage business was regarded as a Jewish occupation, there was some hope that this might be true. As I was then teaching in Czechoslovakia, to which Pressburg belonged at that time, I repeatedly received direct requests from some of the outstanding German physicists to institute inquiries in Pressburg regarding Lenard's four grandparents. I must admit that my interests did not lie in the field of genealogical research. I turned the investigation over to a friend in Pressburg, but he too was not very zealous. The researches did not go beyond Lenard's parents. It was possible to determine that they did not profess the Jewish religion.

Nevertheless, the zeal with which the German physicists had to pursue such problems in the interest of their science is a sign of this peculiar period.

---

5. *Last Weeks in Europe*

Einstein passed the last weeks of his European residence in a villa that lay hidden among the great sand dunes of Le Cocque sur Mer, a beautiful bathing resort in Belgium. Round about, children built large castles of sand and women promenaded in attractive Parisian-model bathing suits. Einstein was in a peculiar situation. He had not returned to Germany, and his friends there warned him that he would certainly be arrested or perhaps murdered if he showed up in that country.

Le Cocque, however, was not very far from Germany. Many feared that fanatics would be able to slip across the border and "liquidate" him. If they fled to Germany after committing such an act, they would not have to fear any punishment, since the deed would have been committed "with the best of intentions." There were several precedents for such actions. It was even rumored that a high price had been set on Einstein's head; but it is manifestly difficult to check the correctness of such talk.

Einstein had good friends in Belgium. The Abbé Lemaître,

a Catholic priest, had found that Einstein's equations of the gravitational field in universal space were also consistent with a distribution of matter in the universe which did not always remain the same on the average. Hence, the Abbé could assume that the various galaxies move farther and farther away from one another. He thus founded the theory of the *expanding universe,* which had been adumbrated in connection with Einstein's theories by Friedmann, a Soviet Russian mathematician more than a decade earlier. It first received attention through Lemaître and still more through Eddington, and was supported by astronomical observations. As Abbé Lemaître was one of the glories of Belgian science, the Queen of Belgium likewise became interested in Einstein's theories and on various occasions took pleasure in conversing with him.

The Belgian royal family and the Belgian government were very much concerned about the rumors that assassins might come to Belgium and threaten Einstein. It was therefore arranged for two bodyguards to watch Einstein day and night. Naturally this was rather annoying for him. In the first place it was unpleasant for a kind-hearted person like Einstein to keep his two shadows too busy, and secondly, for a bohemian like Einstein it was very annoying to be under constant "police supervision." The Belgian government, however, had no desire to be responsible for any accidents.

In the summer of 1933, while passing through Ostend in the course of a trip from London to the Continent, I remembered that Einstein was living near by and decided to try to find him there. I did not know his address, but I took a chance and went to Le Cocque, where I inquired of the inhabitants whether they knew where Einstein was living. As I later learned, the authorities had given strict orders to the inhabitants not to give any information to anyone about Einstein's residence. Since I knew nothing of all these precautions, I asked very naïvely and in time received with equal naïveté all the information that I wanted.

Finally, I came to a villa in the midst of the dunes and saw Mrs. Einstein sitting on the veranda, whereupon I knew I had reached my goal. From the distance I saw two rather robust men in a very excited conversation with Mrs. Einstein. I was rather surprised at these visitors, as one was accustomed to seeing only scientists, writers, and artists with the Einsteins. I approached closer to the villa. As soon as the two men saw me, they threw themselves at me and seized me. Mrs. Einstein

Robert Wood, Planck, and Einstein at a meeting of the German Physical Society, Berlin, 1931, in honor of Michelson

Michelson, Einstein, and Millikan

jumped up, her face frightened and chalky white. Finally she recognized me and said: "They suspected you of being the rumored assassin." She reassured the detectives and led me into the house.

After a while Einstein himself came downstairs. In the meantime Mrs. Einstein had asked me how I had found the house. I replied that the people in the neighborhood had pointed it out to me. But that was strictly prohibited, she said. Einstein himself laughed heartily at the failure of the measures taken by the police for his protection.

At this time his mind was still much occupied by his correspondence with the Academy in Berlin. He showed me all the letters and commented on the parts played by the various persons concerned in the matter. He spoke at some length about the personality of Max Planck. "And finally," he said, "to get rid of my annoyance I composed several humorous verses. I put all the letters in a folder and on top of them the verses. They began with these lines:

> Thank you for your note so tender;
> 'Tis typically German, like its sender."

There was something genuinely artistic in Einstein's nature. It recalled to mind the passage in Goethe's autobiography where he relates that he rid himself of every mental vexation by reenacting it artistically. Einstein in such cases played a short but vigorous composition on the violin or composed a few humorous verses. Even though they did not attain the classical level of Goethe's *Faust,* yet psychologically they fulfilled the same function equally well.

On this occasion Einstein repeatedly emphasized that in getting rid of his Berlin environment he also experienced in a certain respect a psychological liberation. Mrs. Einstein, who was present at this conversation, was not very much in sympathy with such statements. Emotionally she had a strong feeling of attachment to Germany. She said: "But you should not be so unjust. You had many happy hours in Berlin, too. For instance, you often said to me after coming home from the physics colloquium that such a gathering of outstanding physicists is not to be found anywhere else in the world at the present day."

"Yes," said Einstein, "from a purely scientific point of view life in Berlin was often really very nice. Nevertheless, I always had a feeling as if something was pressing on me, and I always had a presentiment that the end would not be good."

We then spoke about the prediction that he had made to me in Prague about eleven years previously, before his first trip to America. The catastrophe in Germany had actually occurred at approximately the time that he had anticipated.

"Do you know," said Einstein, "I have recently had a very remarkable experience. You probably remember my friend and colleague Fritz Haber, the famous chemist." The reader will recall that he belonged to Einstein's intimate circle in Berlin. He had always urged Einstein to adapt himself to the thought of German nationalists, and himself had advanced rather far in this direction. "I recently received a letter from Fritz Haber," related Einstein, "in which he informs me of his intention to apply for a position at the Hebrew University in Jerusalem. There you have it, the whole world is topsyturvy."

We talked a great deal about this university to whose founding Einstein had contributed so much. Now that Einstein had become available, the university in Jerusalem made every effort to obtain his services. But he was not much inclined to accept. He did not like the idea that in this period which was so critical for the Jewish people, the university endeavored chiefly to obtain certain professors who were already famous, in order to increase its prestige. At a time when the future of so many young Jewish scholars was endangered, he felt that this university should rather pick out the most capable of these younger men and enable them to teach and to carry on research. For these reasons Einstein also advised the famous Haber not to go to Jerusalem.

We discussed the fantastic ideas regarding Einstein as a politician that were current among the ruling circles in Germany. Mrs. Einstein related an incident that had occurred recently. They had received a German letter from an unknown man in which he urgently demanded that Einstein receive him. Since no unknown person was permitted to come near him for fear of an assassination, Mrs. Einstein refused. Upon repeated insistence that the matter was very important, Mrs. Einstein finally declared herself ready to see this man in the absence of her husband. The man actually came and related that he had been a member of the Nazi Storm Troops (S.A.). He had fallen out with the party and was now opposed to it. He knew all the secrets of the party and wanted to sell them to its opponents for fifty thousand francs. He wanted to find out whether Einstein would spend the money for this information. "Why do you assume," asked Mrs. Einstein, "that Professor Einstein is interested in the secrets of your former party?" "Oh," replied the ex-S.A. man, "we all know very well that Professor Einstein is the leader of the opposing party throughout the entire world, and that such a purchase would therefore be very important for him." Mrs. Einstein explained to the man that he was mistaken and that Einstein was not interested in these secrets, no matter whether they were genuine or spurious.

Nevertheless, the occurrence left a very uncomfortable feeling. It was now definitely known that the National Socialist Party, which at that time was already one of the most powerful factors in the world, regarded Einstein as a leader of its opponents. All sorts of unpleasant surprises had to be expected.

---

6. *Einstein's Views on Military Service*

Germany's revolution from the Right made it evident to the small neighboring states that the time had come when Germany would break the bonds of the Versailles Treaty, if necessary by force. To any intelligent person acquainted with the lessons of history it was obvious that Germany would not stop with the eradication of the "injustice of Versailles," but would take advantage of the opportunity to obtain something more for herself in order to realize her old dream of a "living space." The war of 1914–18 had made it evident to the Belgians that the German politicians included Belgium within this living space. As early as 1933, at about the time that Einstein came to Belgium, this realization aroused a feeling of insecurity in many persons.

On the other hand, in Belgium as elsewhere at this time, especially among the youth, the view was firmly rooted that all wars are organized by the capitalist class to suppress the workers. Therefore every socially minded and progressive young person should refrain from supporting war in any way. But even then it was already evident to many Belgians that absolute opposition to every war would make the country an easy prey for its neighbors, who preached that war is the most important instrument of politics. Thus radically minded youth was faced by this problem: should the propaganda against military service and military preparedness be continued, thus rendering easier an invasion by warlike neighbors, or should one take part in the defense of the fatherland, thereby following a slogan that had previously been regarded as a pretext of the exploiters in their fight against their own workers? A group of representatives of Belgian pacifist youth turned to Einstein for his opinion in this matter of conscience, since he was widely known as a radical champion of the movement against war and military service. As late as the spring of 1931 he had greeted with delight and affirmation a manifesto issued by American clergymen in which they announced that they would take no part in any future wars, even though their own government claimed that it was for the defense of their country. Einstein had written as follows, referring to this statement:

"It is a gratifying revelation of the temper of the American clergy that fifty-four per cent of those who answered the questionnaire should have indicated their purpose not to participate in any future war. Only such a radical position can be of help to the world, since the government of each nation is bound to present every war as a war of defense."

But when the young Belgians turned to Einstein with the question whether they should refuse to co-operate if Belgium became involved in a war against its big neighbors, Einstein did not let himself be confused for a moment. From the very first he knew that he had to answer in such a way as to encourage the course of action that he considered advisable under the given circumstances. He did not allow himself to be confused by the vain idea of standing forth as one who sticks to his principles under all circumstances. Such a person would insist on his principles even though they should lead to actions and results with which he was not in sympathy. Einstein was aware that the purpose of principles in both public and private life is only to encourage actions that produce results which one would approve. Principles, however, are not to be considered as ends in themselves. He answered briefly and concisely: in this case everyone should fight as best he can for the freedom of his fatherland, Belgium.

This answer created a sensation at the time. Many persons even doubted its genuineness. Many said: "Surely a principle does not become false because in a single case it leads to consequences that are repugnant, as for instance in this case to a triumph of National Socialism."

The people who expected that Einstein would stick to his principles without any consideration of the consequences did not understand the fundamentally positivistic, pragmatic character of his thought. Basically he thought in politics just as he did in physics. When he actually came to grips with a concrete problem, the positivistic basis of his thinking became evident. He did not believe that principles have any meaning except their consequences, which we can test on the basis of our experiences. Occasionally, he liked to think about the emotional effect brought about by the wording of the principles. As a result his language in physics as well as in politics in some cases acquired a metaphysical touch. But this was only a more or less poetical way of speaking, which furnished a point of contact with human feeling.

Basically his position was always clear: he would never support principles because of their beautiful sound, if they led to consequences which he could not approve.

For this reason the attacks on Einstein by those who opposed war on principle were of the same character as those of some of his opponents in physics who attacked him with the reproach that he had first advanced the principle of the constancy of the velocity of light in the special relativity theory of 1905, and had then abandoned it in his theory of gravitation, since according to the latter the velocity with which light is propagated depends on the intensity of the gravitational field. Some of Einstein's opponents accused him of being inconsistent and of trying to hide this inconsistency. This description, however, is somewhat misleading. The constancy of the velocity of light is true only under very specific conditions — namely, when strong gravitational fields are not present. By enumerating the restrictions under which a certain principle is valid, one is not being inconsistent, but only adding to our knowledge of the world.

The same is true of Einstein's attitude to the question of military service. At that time I had no opportunity to discuss this matter personally with him, but soon after Einstein's arrival in America the same question became acute there. The radical youth movement, as represented by the American Youth Congress, at first wanted to uphold the principle of absolute opposition to war, even in the case of a war of the democratic states against fascism, because such a war for them was in principle an imperialist war. Einstein, however, did not let himself be confused by such arguments, and saw that here as in Belgium these "opponents of war" were only working for the victory of the greatest military power. As a result they would achieve the very opposite of what they thought they were working for. Einstein thought that the principle of absolute non-participation in war made sense only when a victory of the different powers did not lead to very different consequences for the population. In Europe after 1918 one might have said: It does not make much difference whether one is ruled by the French or the German Republic, by the United States or Great Britain. This difference does not justify war. But this standpoint can no longer be maintained when there are states whose principles of government differ as radically from each other as do those of Nazi Germany from those of the states around it. Under these conditions no one can remain indifferent to who will be the victor. Just as the principle of the constancy of the velocity of light is valid only if no great differences of gravitational potential and therefore no great forces are present, so the principle of absolute refusal to perform military service is valid only when there are no extreme differences between the governmental principles of opposing states.

In U.S.A. opponents of military service such as Bertrand Russell and Archibald MacLeish drew the same consequences from the situation. Various metaphysically thinking authors characterized such men as "inconsistent" and wondered that logicians such as Russell could be so illogical. Einstein's case, however, would already have shown them that consistency in a metaphysical sense — that is, to hold fast to the letter of a principle — is not consistency in a scientific sense which means to hold fast to the desirable consequences of a principle. Thus, because of his direct and honest thinking, Einstein once again became an object of attacks, even before he had actually departed from Europe, and this time the attacks came from "progressive" and "radical" circles.

At this time Einstein was most immediately concerned with the many hundreds and soon thousands of scholars and scientists, both young and old, who were expelled from their positions by the purge in Germany. English scientists tried to give the refugees some opportunity to continue their work under more favorable conditions. The great English physicist Rutherford put himself at the head of this movement and organized the Academic Assistance Council in London. At its first meeting Einstein was to be presented to the public as a symbol of the victims, and with his great prestige was to make an appeal for this cause. One can very well imagine that this was not very pleasant for Einstein. He did not like to appear publicly in any matter where he was personally involved. Nevertheless, the seriousness of the situation and the importance of the relief measures

induced him to go to London and deliver an address on the subject "Science and Liberty." At the meeting he sat next to Lord
Rutherford, who presided. Immediately after his introductory
words Rutherford pointed to his neighbor with an energetic
gesture and presented him proudly: "Ladies and gentlemen, my
old friend and colleague Professor Einstein."

Einstein spoke with great reserve. He tried to point out the
need for relief measures, while avoiding all political attacks.
Strong words were superfluous, the cause spoke for itself. Einstein said: "It cannot be my task to act as judge of the conduct of a
nation which for many years has considered me as her son. Perhaps it is an idle task to judge in times when action counts."

Soon after this meeting, which took place early in October
1933, Einstein was waiting at Southampton for a passenger vessel of moderate size that was coming from Antwerp and was
to bring him to New York.

But before I describe Einstein's new life in America, we will
remain awhile yet in Europe to see the remarkable manner in
which Einstein's abstract theories were utilized by political and
religious groups for their purposes.

XI

EINSTEIN'S THEORIES AS POLITICAL

WEAPONS AND TARGETS

1. *Scientific Theories and Political Ideologies*

To a physicist or mathematician who actually understands, or believes that he understands, Einstein's theories, it must seem strange and frivolous when people whose understanding of this matter is much more limited argue whether his theory is a product of the Bolshevization of Europe or perhaps a stage in the development of Europe from liberalism to fascism; whether it is a support for religion in its fight against materialism or whether it helps to breed disbelief in everything that traditional religion teaches about the universe. The professional physicist will not find any trace of these ideas in Einstein's theories. He believes that their validity depends only on the correctness of certain computations, and on whether certain delicate experiments are carried out with the necessary care. Consequently he must feel that these disputes over Einstein's theories have been simply a result of ignorance and madness.

But whoever investigates the fate of other radically new theories about the universe — for example, the fate of the Copernican system, the Newtonian theory, the laws of energy — will find that all these theories led to discussions that from the standpoint of the physicist or mathematician appeared to be either superfluous or even foolish.

The transition from science to political ideology occurs by means of philosophy. The generalizations of science are expressed in philosophical language, in which terms such as "idealism," "materialism," "force," "energy," and others play a part. The same words also appear in the philosophical doctrines that tell men how to act in private as well as in political life. In this way the generalizations of science are gradually transformed into principles of moral and political philosophy.

On this point Viscount Samuel, a man who is conversant with

science, philosophy and politics, and who in addition has been connected with Einstein in a number of ways, said:

"Philosophy of some kind moves the nations. Every land resounds with the tramp of armies, behind the armies are the dictators and the parliaments, behind them are the political creeds — Communism, National Socialism, Fascism, Democracy — and behind the creeds are the philosophers — Marx, Engels, Hegel, Nietzsche, Sorel, Mill, and others."

Philosophical systems like to make use of the newest scientific theories in order to have "exact" foundations. But the help that philosophy gets in this way does not lead to unambiguous results. One and the same scientific theory can be used to support different political creeds. Bertrand Russell gave a very good characterization of this ambiguity:

"There has been a tendency, not uncommon in the case of a new scientific theory, for every philosopher to interpret the work of Einstein in accordance with his own metaphysical system and to suggest that the outcome is a great accession of strength to the views which the philosopher in question previously held."

This ambiguity arises from the fact that it is not the physical content of a theory that is responsible for its philosophical interpretations. Frequently it is rather the language in which the theory is formulated, its images and analogies that are interpreted.

The interpretation of Einstein's Relativity is usually connected with two characteristics of the language in which he and his followers clothed his theory. The first characteristic is the abandonment of mechanical analogies. There is no mention of any mechanism in the sense in which this word is used in daily life; for instance, there is no mechanism for the shortening of a body by rapid motion. Instead, a *logical-empirical* mode of expression is employed; that is, a system of mathematical formulæ is given and the operations are described by which the magnitudes in these formulæ can be measured empirically. The second characteristic is the use of the expression "relative to a certain body." The use of this mode of expression gives rise to a comparison with the language of so-called "relativism"; for example, *ethical relativism,* which asserts that any human action can be called good or bad only "relative to a certain ethnical group and historical period," and so forth.

By abandoning the mechanical analogy Einstein's theory harmonized to a certain extent with all the currents of thought that opposed the mechanistic conception of the world and the materialistic philosophy connected with it. The second characteristic of his mode of expression brought him close to those who were called ethical skeptics and who were frequently linked to a materialistic philosophy.

Thus Einstein's theories could be used equally well as propaganda for materialism or against it. And since words such as "materialism," "idealism," "relativism," and so forth, are frequently used as the catchwords of political ideologies, we can understand that Einstein's theories were very often used as a weapon in the struggle of political parties.

2. Pro-Fascist Interpretation

The fascist groups always have asserted that the Communist philosophy is materialistic, while theirs is anti-materialistic, or idealistic. Consequently Einstein's theories could be used as weapons for fascism if they were interpreted as arguments against materialism and for idealism.

As early as 1927 — that is, before the seizure of power by the Nazis — Joseph Goebbels had shown how the language of German idealistic philosophy could be employed in the service of his party. First of all he presented an interpretation of the Kantian expression "thing-in-itself" (*Ding an sich*), the characteristic concept of German idealism. Goebbels said: "The folk is a constituent of humanity. Humanity is not a thing-in-itself, nor is the individual a thing-in-itself. The folk is the thing-in-itself. . . ."

"The materialist," Goebbels continued, "regards the folk only as an instrument and does not want to concede that it is an independent objective reality. For him the folk is an intermediate thing between man and humanity, and mankind is for him the ultimate. . . . Therefore the materialist is necessarily a democrat. The idealist sees in the word 'humanity' only a concept. Humanity is only something imagined, not a fact. . . ."

By emphasizing its anti-mechanistic aspects it was, indeed, possible to employ Einstein's relativity theory as a weapon in the fight against "materialistic" democracy. German physicists who considered it desirable to teach Einstein's theories even in National Socialist Germany occasionally made use of this possibility. Pascual Jordan, for instance, in his book *The Physics of the Twentieth Century,* recommended Einstein's theory of relativity to National Socialists as a weapon in the fight against materialistic philosophy. Jordan said that the eradication of this philosophy is an "integral aspect of the unfolding new world of the twentieth century that has already begun, especially in Italy and Germany." The "new world" is that of Fascism and National Socialism.

Since many opponents of Einstein's theories wanted to make use of the political power of the National Socialist Party in their fight against Einstein, they were very much upset by efforts such as that of Jordan. Thus, for instance, Hugo Dingler, who had already agitated against Einstein without any great success long before National Socialism, remarked with indignation about Jordan's book: "To hang this destructive Einstein philosophy on the skirts of the national movements in Germany and Italy is really a little too much."

With the adjective "destructive" Dingler touched directly upon the other feature in the language of the theory of relativity, the use of the expression "relative." He connected Einstein's theories with the English philosophy of enlightenment of David Hume, which according to popular conception is only a variant of materialism, and which the National Socialist Party felt obliged to oppose.

If the theory of relativity had been advanced by someone other than Einstein, it is entirely possible that it would not have been unanimously condemned by the National Socialist Party. The relativity theory would very possibly have remained a constant object of controversy in these circles like various other philosophies. Einstein's Jewish ancestry, however, and his political attitude as a pacifist made the condemnation of his theory inevitable.

3. *Einstein's Theories Attacked as Expressions of Jewish Mentality*

In general, National Socialist writers regarded two groups of characteristics as typical of Jewish thinking. In the first place, it was said, the Jew prefers pure speculation to experimental observations of nature. Secondly, it was asserted that the Jew does not recognize purely mental concepts, but believes only in truths that can be discovered by sensory experience of material things. Obviously it is not difficult to find one of these characteristics in any physicist.

Among those who attacked Einstein on the ground that his theories were purely speculative, the most ardent was Philipp Lenard, who has been mentioned several times already. In his book *German Physics* he said:

"Jewish physics can best and most justly be characterized by recalling the activity of one who is probably its most prominent representative, the pure-blooded Jew Albert Einstein. His relativity theory was to transform and dominate all physics; but when faced with reality, it no longer has a leg to stand on. Nor was it intended to be true. In contrast to the equally intractable and solicitous desire for truth of the Aryan scientist, the Jew lacks to a striking degree any comprehension of truth — that is, of anything more than an apparent agreement with a reality that occurs independently of human thought."

In a lecture delivered at Munich in 1937 before the association of provincial teachers and students (*Gaudozentenbund und Gaustudentenbund*), the origin and development of this "Jewish" way of looking at nature was related to political conditions after the first World War. It was said:

"The entire development of natural science is a communal effort of Aryan scientists, among whom the Germans are numerically foremost. The period of Heinrich Hertz coincides with the gradual development of a Jewish natural science, which took advantage of the obscure situation in the physics of the ether and branched off from the course of development of Aryan physics. By systematically filling academic positions with Jews and by assuming an increasingly dictatorial attitude, this Jewish natural science tried to deprive Aryan physics of its foundations, to dogmatize, and to oppress all thinking about nature. Ultimately it replaced these foundations by a deceptive imaginary structure known as the relativity theory, above which it simultaneously inscribed the typically Jewish taboo — that is, 'not to be touched.' This development was temporarily and causally coincident with the victory of Jewry in other fields during the postwar period."

In 1938 the *Zeitschrift für die gesamte Naturwissenschaft* (*Journal for General Science*) was founded for the specific purpose of propagating the National Socialist conception in science. In an article: "Racial Dependence of Mathematics and Physics," we read the following:

"The influence of the Jews on the development of natural science is due first of all to a difference in their attitude toward the fundamental relation between experiment and theory in favor of the latter. Theories were constructed without regard for the forms of human thought and perception and without any rigorous methodology of reasoning. . . . Einstein's theory of relativity offers us the clearest example of a dogmatic Jewish type of theory. It is headed by a dogma, the principle of the constancy of the velocity of light. In a vacuum the velocity of light is supposed to have constant magnitude independent of the state of motion of the light sources and the observer. It is falsely asserted that this is a fact of experience."

Actually Einstein's principle of the constancy of the velocity of light is just as much and just as little a fact of experience and a dogma as any other of the basic hypotheses of a physical theory. It is only because of erroneous and defective presentations of Einstein's theory that many persons believe the relation between theory and experience to be different here from what it was in the older theories.

This alleged preference of the Jews for theoretical deliberations was contrasted with the striving of the Aryan German for concrete action. The same contrast was seen in politics: the eternal pondering and indecision of the democratic states; and the firm action of National Socialist Germany.

But by the average spokesmen of Nazi philosophy Einstein's theories were branded as materialistic and thus linked with Marxism. In 1936 a lecture was given at the camp of the Natural Scientific Professional Group of the National Socialist Student Association, in which it was said:

"Einstein's theories could only have been greeted so joyfully by a generation that had already been raised and trained in materialistic modes of thought. On this account it would likewise have been unable to flourish in this way anywhere else but in the soil of Marxism, of which it is the scientific expression, just as this is true of cubism in the plastic arts and of the melodic and rhythmic barrenness of music in recent years."

The speaker summarized his views in the statement: "The formulation of general relativity as a principle of nature cannot be anything but the expression of a thoroughly materialistic mental and spiritual attitude."

Comparisons can certainly be drawn between the expressions of a period in different fields. But that Einstein's theories had developed on the basis of Marxism has certainly not been evident to the Marxists, as we shall soon see.

The same speaker later (1937) commented on his remarks, saying:

"Under the influence of the philosophy of enlightenment the nineteenth century was a period that was excessively attached to the surface of things and valued material things beyond all measure. Hence, the majority of scientists were unable to grasp and to develop the concept of the ether which by its very nature obeys other laws than those of matter. Only a few, among them Philipp Lenard, had the breadth of soul and mind that was necessary for such a step. The others fell into the hand of the Jew, who instinctively grasped and exploited the situation."

In order to be able to judge these arguments, one must remember that the ether was introduced into physics only to explain phenomena by analogy with mechanics. Einstein was the first to recognize the impossibility of a mechanical explanation of optical phenomena, and therefore got rid of the ether. This was the consistent action of a man who recognized the untenableness of the mechanistic conception of nature. The scientific supporters of the National Socialist Party did not want to take this step. They did not want to give up the mechanistic conception of physics, since it somehow fitted into their philosophy of unsophisticated approach to nature. But as they were simultaneously opposed to materialism, their position became a rather difficult one. They introduced an ether that was not material, and thus had none of the properties for the sake of which it was introduced.

Later Lenard also proposed this compromise solution. Since the seizure of power by the Nazis, he attacked Einstein from a new standpoint. Previously he had opposed Einstein because the latter had given up mechanistic explanations in physics; now he accused Einstein of materialism and failing to recognize an immaterial ether. Einstein, however, introduced no mechanical basis whatever for optical phenomena and was farther removed than Lenard from materialism in this mechanistic sense.

Another reason for the opposition to Einstein came from the circumstance that the word "force" is a term that was used with particular favor by the National Socialists. They considered it a great misfortune that this word should disappear from physics. The fight for this word reveals very clearly the manner in which physics and politics are connected.

The Austrian Ernst Mach and the German Gustav Kirchhoff were the first among the physicists to construct a system of mechanics in which the word "force" did not occur in the laws of motion. This word was introduced only as an auxiliary concept to abbreviate the mode of expression. Since the National Socialists characterized everything that they did not like as "Jewish," they regarded the elimination of the word "force" as the work of the Jews, even though, as we have seen, it was undoubtedly first carried out by German physicists. In his *Mechanics* Heinrich Hertz, the discoverer of electrical waves, followed Mach and Kirchhoff in seeking a new way to eliminate the word "force" from the fundamental laws of motion. National Socialist authors ascribed this striving to Hertz's Jewish blood. One of them writes: "If we recall that the Jewish physicist Einstein also wanted to remove the concept of force from physics, we must raise the question at this point whether an inner, racially determined relationship does not appear here." In Einstein's theory of gravitation the concept of force does not appear as a basic concept. Bodies move in paths that are represented by the "shortest" possible curves.

This elimination of force as a fundamental concept is regarded as characteristic of the Jewish type of thinking. In an article in the *Zeitschrift für die gesamte Naturwissenschaft* we read:

"The concept of force, which was introduced by Aryan scientists for the causal interpretation of changes in velocity, obviously arises from the personal experience of human labor, of manual creation, which has been and is the essential content of the life of Aryan man. The picture of the world that thus arose possessed in every detail the quality of visual clarity, from which arises the happy impression that it produces on related minds. All this changed fundamentally when the Jew seized the reins in natural science to an ever increasing degree. ... The Jew would not be himself if the characteristic feature of his attitude, just as everywhere else in science, were not the disintegration and destruction of Aryan construction."

The author links "Jewish Physics" with a favorite Nazi target, the *Talmud:*

"The mode of thought that finds its expression in Einstein's theory is known, when applied to other ordinary things, as 'Talmudic thinking.' The task of the Talmud is to fulfill the precepts of the Tora, the Biblical law, by circumventing them. This is accomplished by means of suitable definitions of the concepts occurring in the law and by a purely formalistic mode of interpreting and applying them. Consider the Talmud Jew who places a food receptacle under his seat in a railway car, thus turning it formally into his residence, and in this manner formally obeying the law that on the Sabbath one should not travel more than a mile from his residence. It is this formal fulfillment that is important for the Jews.

"This formalistic Talmudic thinking likewise manifests itself in Jewish physics. Within the theory of relativity the principle of the constancy of the velocity of light and the principle of the general relativity of natural phenomena represent the 'Tora,' which must be fulfilled under all circumstances. For this fulfillment an extensive mathematical apparatus is necessary; and just as previously the concept of 'residence' and 'carrying' were rendered lifeless and given a more expedient definition, so in the Jewish relativity theory the concepts of space and time are deprived of all spirit and defined in an expedient, purely intellectual way."

The characterization of the Einsteinian definitions of "length," "temporal duration," and so on, as "lifeless" in contrast to the definitions of traditional physics has only the following justification. At every stage of scientific development concepts are introduced by means of such definitions as correspond to the particular stage; that is, they are as practical as possible for the presentation of available knowledge. When such a stage has lasted for a long time, the words that are used in science gradually become words of daily life; they acquire an emotional overtone and become filled with life. Every introduction of new definitions appears to us to create "lifeless" concepts.

I once met on a train a Japanese diplomat who was just coming from the Wagner festival at Bayreuth. I asked him how he liked Wagnerian music. He replied: "Technically it is highly developed and ingenious. But in comparison with Japanese music it lacks a soul." For one who has grown up with the sound of Japanese music in his ears Wagnerian music sounds just as "lifeless" and "intellectualistic" as the definitions of Einstein's theory do to one who has been accustomed all his life to Newtonian mechanics.

4. *Attitude of the Soviet Philosophy toward Einstein*

The Soviet government publishes the *great Soviet Encyclopædia* that presents the entire knowledge of our time from the point of view of Soviet doctrines. This article on "Einstein" begins with the words: "Einstein is the greatest physicist of our time." Among the Soviet philosophers he is regarded as a great physicist who was prevented by the economic circumstances under which he grew up from drawing the correct philosophical conclusions from his theory. Regarding Einstein's philosophical views the *Encyclopædia* reads: "Einstein's philosophical position is not consistent. Materialistic and dialectical elements are interwoven with Machist assertions, which predominate in almost all of Einstein's remarks."

In order to understand these comments, it must be remembered that *dialectical materialism* has been the official philosophy of Soviet Russia, and that *Machism,* the teaching of Ernst Mach, has been the main target of its attack.

On consulting the article on "Ether" in the same *Encyclopædia,* we find there:

"In physics we often find a completely erroneous contrast between ether and matter. Since the physicists regard only gravity and inertia as criteria of materiality, they were inclined to deny the materiality of the ether. Here we have the same confusion of the physical and philosophical concepts of matter that was analyzed by Lenin in his consideration of the crisis in natural science at the beginning of the twentieth century. . . . The ether is a kind of matter and has the same objective reality as other kinds. . . . The antithesis of ether and matter is senseless and leads to agnostic and idealistic arguments. . . . The theory of relativity has recourse to a mathematical description, it abandons the answering of the question concerning the objective nature of physical phenomena; that is, in the question of the ether it takes the standpoint of Mach."

By studying events in Russia since the seizure of power by Lenin, we can see that no attempt was ever made to exert political influences on physical theories proper, and when individuals did attempt to do this, it was not approved by the authorities. On the other hand, the philosophical interpretation of theories has been a political matter; the intervention of the party and its organs, as for example, the Communist Academy of Science in Moscow, has been regarded as a matter of course. Naturally, the boundary between a physical theory and its philosophical interpretation cannot always be drawn so distinctly, and on various occasions border trespassing has taken place. Lenin had already said on one occasion: "Not a single professor among those who are able to make the most valuable contributions to the special domains of chemistry, history, or physics, can be trusted even so far as a single word when it comes to philosophy."

The official conception of the reciprocal relations between

physics, philosophy, and politics is very clearly set forth in an address delivered by A. F. Joffe, the leading physicist of the Soviet Union, in 1934 at a memorial session of the Philosophical Institute of the Communist Party. This session was held to commemorate the publication twenty-five years previously of Lenin's chief philosophical work, *Materialism and Empiriocriticism,* which contains Lenin's views on the misinterpretations of modern physics and his attacks on "Machism." In his address Joffe said:

"When physicists such as Bohr, Schrödinger, and Heisenberg express their opinions in popular works regarding the philosophical generalizations of their work in physics, their philosophy is sometimes a product of the social conditions under which they live and of the social tasks that they carry out, either consciously or unconsciously. Thus Heisenberg's physical theory is a materialistic theory; that is, it is the closest approximation to reality possible at present. Lenin, too, did not criticize Mach's scientific researches, but only his philosophy."

The philosophers of the Roman Church made, already, a clear distinction between the astronomical theories of Copernicus and Galileo's philosophical interpretation of these theories.

In 1938 A. Maximov, one of the best-known Soviet writers on the philosophy of physics, said in an article on the significance of Lenin's book mentioned above:

"No physical theory has produced such a stream of idealistic fantasies as Einstein's theory of relativity. Mystics, clerics, idealists of all shades, among them also a number of serious scientists, snatched at the philosophical consequences of the theory of relativity. The idealists directed all their efforts to the refutation of materialism. Somehow it had proved the philosophical relativity of time and space. Then came the general theory of relativity, together with the theory of the curvature and finiteness of space."

By the phrase "refutation of materialism" is here meant the proof that departures from Newtonian mechanics and the ether theory of light are necessary. Maximov then referred explicitly to the political causes of the idealistic tendencies manifested by the scientists. He said:

"In our time the bourgeoisie in a number of countries has abandoned the veiled forms of capitalistic dictatorship for the open dictatorship of ax and bludgeon. As a result of the persecutions of the scientific *Weltanschauung* in the capitalistic countries, which was connected with this transition, many scientists joined the camp of reaction. This change of loyalties manifests itself among scientists by the appearance of avowals of idealism and metaphysics. During the last ten to fifteen years a retrograde trend has manifested itself in all fields of natural science in the capitalistic countries. Opposition to Darwinism and to the Kant-Laplace theories in physics, and attacks on the law of the conservation and transformation of energy, have become the fashion."

It is certainly true that idealistic "interpretations" of the relativity theory have been frequently used to bolster up Fascist philosophy.

Soon after the general theory of relativity achieved world fame, in 1928, this same Maximov described it as a plant that had sprouted in the mystically inclined soil of the postwar period. After describing the postwar years in Germany, he said:

"This idealistic atmosphere surrounded the theory of relativity and still surrounds it at the present day. It is therefore only natural that the announcement of general relativity by Einstein was received with delight by the bourgeois intellectuals. The inability of scholars to withdraw from this influence within the boundaries of bourgeois society led to the result that the relativity principle served exclusively religious and metaphysical sentiments.

"What should be our relation to the theory of relativity? We should accept all the empirical material as well as all the conclusions and generalizations that follow logically from it. . . . But in place of the idealistic presentation of the theory of relativity favored by bourgeois society we must develop a dialectical presentation of the theory. We need young capable scientists who are thoroughly imbued with the proletarian ideology."

In order to understand correctly the Soviet attitude toward Relativity we have to distinguish two periods. During the first years of the Soviet regime there prevailed among the official philosophers the opinions that the relativity theory contradicted materialism because it did not regard optical phenomena as phenomena of motion occurring in a material body. This view was supported by the Moscow physicist A. K. Timiryasev, who judged all physicists on the basis of their agreement or nonagreement with Newton's mechanistic science.

It will be recalled that Lenard, the leading Nazi physicist in Germany, had also rejected Einstein's theory because it could not present a mechanical model of optical events. Soon after its publication in 1922, Lenard's book was translated into Russian and published with an introduction by Timiryasev. In the same year Maximov wrote a review of Lenard's book for the leading philosophical journal of the Soviet Union, *Under the Banner of Marxism,* in which he said:

"While Einstein, the idealist, ascribes an absolute value to the creations of the mind and puts the world of events on an equal footing with the world of experiences, Lenard takes a diametrically opposed point of view. From the standpoint of common sense, which is more inclined to stick to the experiences of the material world than to the need of philosophy, Lenard prefers to retain the mechanical picture of the world. Starting from a standpoint that is in general purely materialistic, Lenard clearly recognizes the contradiction to which one is led by the theory of relativity."

On the other hand we have seen that spokesmen of Nazi philosophy frequently asserted that Einstein's theory could have flowered only in the soil of materialism and that it appears together with Marxism. Now we see that the authorized interpreters of Marxism were apparently not of this opinion. We also see that the description of a physical theory as "materialistic" or "idealistic" depends only upon its philosophical interpretation.

The attacks of the earlier Soviet philosophers upon Mach and Einstein coincided at many points with those of the National Socialist writers. We need only consider the criticisms that Einstein's theories only "describe" nature, but do not "explain" it, that they reject everything that cannot be an object of sensory experience, that they lead to general skepticism and to the destruction of all objective knowledge of nature, and so forth.

Later on confusion of materialism with "mechanistic physics" has been denounced by the "Soviet Institute of Philosophy" as a "reactionary" doctrine that is not compatible with modern science. By "materialism" one should not mean that all natural phenomena could be reduced to motions following Newton's law. This "mechanistic materialism," as a matter of fact, had been denounced already by Marx and Engels. But it enjoyed a comeback, as some physicists used it as a weapon against Einstein's theory as Nazi physicists like Lenard had done in Germany. In stressing *dialectical materialism* in the sense of Marx, Engels, and Lenin, "materialism" means that science has to do with objective facts that are independent of human consciousness; but these facts do not need to be merely motions of material particles.

In the second period of Soviet philosophy, after the abandonment of "mechanistic" materialism, a leading Russian physicist,
Vavilov, demonstrated that the theory of relativity is quite in
agreement with materialism if this word is interpreted in the
sense of Marx, Engels, and Lenin. In an article that appeared in
1939 Vavilov said clearly:

"Objective real space devoid of material properties, motion divorced
from matter, are metaphysical phantoms that sooner or later have to
be expelled from the physical picture of the world. . . . The historic
service rendered by Einstein was the criticism of the old metaphysical
conceptions of space and time. . . . In Einstein's theory space-time
is an inseparable property of matter itself. Such is the basic idea of
Einstein's general theory of relativity. The idealistic conception of
space-time as a category of thought is swept away. . . . Before us is
the first outline, still far from perfect, of the dialectical materialistic
understanding of space and time. Once again dialectical materialism
has triumphed."

More recently, the danger of a "pure philosophy" separated
from science has been more and more recognized in the Soviet
Union. A close co-operation between scientists and philosophers
has been more and more required as the only basis of progressive
thought. Discussions between physicists and philosophers removed the most harmful misunderstandings, and in 1942, after
"twenty-five years of philosophy in the U.S.S.R." the leading
Soviet philosopher, the Academician M. Mitin, gave an address
in the Russian Academy of Science at the twenty-fifth anniversary of the Soviet Union in which he celebrated as an important achievement of these twenty-five years of philosophy
the fact that the attacks against Einstein's theory had ceased and
its compatibility with a sound brand of materialism had been
established.

"As a result," says Mitin, "of the tremendous work that our
philosophers and physicists had carried out, as a result of many im-
passioned discussions . . . it may be now said that our philosophical
conclusions concerning this theory have been firmly established. The
theory of relativity does not deny that time and space, matter and
movement, are absolute in the sense of their objective existence outside human consciousness. . . . The theory of relativity establishes
only the relativity of the results of measuring time and space by observers who are moving relatively to one another."

And then Mitin proceeds to characterize Einstein's theory in
almost the same words that Einstein himself had used to sum-

marize the gist of his theory in one sentence to the newspapermen who interviewed him at his first arrival in New York
Harbor.

Mitin says:

"Time and space are indivisible from the moving body and must be
regarded relative to that movement. In this respect time and space
are relative. . . . In place of the old metaphysical conception of pure
time and space having only geometrical qualities, we obtain a new
theory of time and space inseparable, bound up with bodies and
movement."

---

5. *Einstein's Theories as Arguments for Religion*

We have seen how Einstein's theories were linked to
expressions such as "materialism" and "idealism" in a fairly ambiguous manner and in this way used to support political creeds.
It is not surprising therefore that they were used in a similar
fashion in the battle over religious ideas.

It will be remembered how (ch. VIII, sect. 6) the Archbishop
of Canterbury had gone to a great deal of trouble to study the
theory of relativity, and how he had felt reassured by Einstein's
statement that this theory had nothing to do with religion.
Nevertheless, a man like Sir Arthur Eddington, who not only
was an outstanding astronomer and thoroughly conversant with
the theory of relativity, but had also achieved a great reputation
in the field of the philosophy of science, did not agree at all with
Einstein's remark. In his book *The Philosophy of Physical Science,* published in 1939, he said that Einstein's answer to the
Archbishop was not very conclusive.

Consequently I wish to describe some attempts that have been
made to establish a connection between Einstein's theories and
religion. Once again the course taken was by way of philosophy,
and here too the starting-point was the question: Is Einstein's
theory idealistic or materialistic?

Several years ago in an address to Catholic students Cardinal
O'Connell, Archbishop of Boston, said:

"Remembering the tremendous excitement over the Darwinian
theory of evolution during my boyhood and the furore created less
than ten years ago by Einstein's theory of relativity I tell you that those
theories became outmoded because they were mainly materialistic
and therefore unable to stand the test of time."

Nevertheless, Catholic philosophers were themselves not agreed whether Einstein's theory is actually materialistic. The Irish philosopher A. O'Rahilly, who is also thoroughly conversant with theoretical physics, disagrees with Einstein's theory of relativity because it is based on "subjective idealism."

Thomistic philosophy, which is at present generally regarded as the scientific foundation of Catholic theology, rejects both idealism and materialism. Consequently, to the Catholic who takes his stand on the basis of Scholasticism, either philosophical interpretation of Einstein's theory is a weapon that can be turned against him. If, however, the scholastic foundations of religion are not considered and one consults simply one's feeling, then a religious person will regard any theory that can be interpreted as an argument for idealism as supporting his faith. On the occasion of Einstein's visit to London the conservative *Times* had stated triumphantly in an editorial: "Observational science has, in fact, led back to the purest subjective idealism."

What the journalist stated briefly and concisely for the public at large was soon demonstrated professionally by the British philosopher Wildon Carr in a book for philosophers and theologians. In it he said:

"The adoption of the principle of relativity means that the subjective factor, inseparable from knowledge in the very concept of it, must enter positively into physical science. . . . Hitherto the scientific problem has been to find a place for mind in the objective system of nature and the philosophical problem to validate the obstinate objectivity of nature. . . . Now when the reality is taken in the concrete, as the general principle of relativity requires us to take it, we do not separate the observer from what he observes, the mind from its object, and then dispute as to the primacy of the one over the other."

According to this, the achievement of the relativity theory for religion is simply that it provided a place for mind in nature, which during the period of mechanistic physics had been regarded as completely "material and mindless."

If the reader will recall Einstein's physical theories, he will easily see that this interpretation is more closely related to the wording than to the content of these theories. This is even more obvious in the case of authors who use the four-dimensional representation of the theory of relativity as an argument for traditional religion. As a typical example I wish to quote from an article written by the director of the Department of Theology of an English college abroad, which appeared in the *Hibbert Journal* in 1939. He said there:

"If the idea of time as a fourth dimension is valid, then the difference between this mortal life and the 'other life' is not a difference in the time nor in the quality of the life. It is only a difference in our view of it — our ability to see it whole. While we are limited to three-dimensional understanding, it is mortal life. Where we perceive it in four dimensions, it is eternal life."

This is obviously an interpretation of the words used in the theory of relativity and has hardly anything to do with its factual content. Einstein's own attitude to religion, however, has never been determined by his particular physical theories, but rather by his general judgment about the role of science and faith in human life. The numerous attempts to make the theory of relativity a springboard for excursions into the field of theology have never been encouraged by Einstein.

XII

EINSTEIN IN THE UNITED STATES

1. *The Institute for Advanced Study*

As the racial and political purging proceeded in the German universities, it soon became evident throughout the entire world that a large number of capable and often famous men were looking for positions outside Germany. It thus became possible for institutions abroad to acquire many outstanding scholars cheaply. One of the greatest German scientists, whom I visited at his Berlin laboratory in the summer of 1933, showed me a long list of men who were available and said half-jokingly: "What we are now doing in Germany is organizing a bargain sale of good merchandise at reduced prices. Shrewd persons will certainly seize this opportunity to buy something from us."

The scholars who had been dismissed in Germany could actually be compared in this way to merchandise that had to be sold at reduced prices as "irregulars." Even such a slight defect as the ancestry of a scientist's wife made the sale necessary. And of these bargains which appeared on the market at that time, Einstein naturally created the most sensation. It was as if a great museum were suddenly to offer for sale Rembrandt's most valuable paintings at a very low price simply because the new directors of the museum did not like to have pictures of a certain style.

Einstein, of course, did not have any troubles in finding a new position. Many universities offered him posts. The Universities of Madrid and Jerusalem, among others, invited him, and one of the oldest and most esteemed institutions of Europe, the venerable Sorbonne in Paris, actually appointed Einstein a professor though he never really occupied this position. Einstein wanted to leave Europe because he did not expect any change for the better in the immediate future. His friends also cautioned him against settling anywhere near Germany since, in view of the fantastic ideas regarding Einstein's political influence and activity held by the ruling party, the danger was always present that some fanatic might order Einstein to be "liquidated."

Einstein had no difficult decision to make, since he had been offered already and had accepted an ideal position in the United States. The offer had been made in the summer of 1932, and at that time it was for him an unexpected sign from heaven to prepare for emigration from Europe.

In 1930 Mr. Louis Bamberger and Mrs. Felix Fuld, on the advice of Abraham Flexner, who had done so much for the reform of American education, donated a sum of five million dollars for the founding of an entirely novel institution for research and teaching. They had asked Dr. Flexner how in his opinion they could most usefully employ their money, and he had replied that there were already in the United States many universities where students could work for the degree of Doctor of Philosophy, but he felt the lack of another type of institution. He had recognized the important need for promising young scholars who had completed the work for the doctorate to continue their training and research in daily informal intercourse with the leaders in their fields. Flexner felt that this informal contact between outstanding scholars and students had been the great achievement of the German universities during their golden era. In his opinion the American universities were still not yet adequately organized for this purpose, with the courses only serving to prepare students for certain academic degrees and the professors too greatly overburdened to maintain any contact with students who had completed their studies.

This institute, which was named the *Institute for Advanced Study,* and whose direction Dr. Flexner was asked to assume, was to be an institution in which a small group of professors served as the nucleus of a larger, temporary group of mature, though generally younger scholars. The choice of the staff and admission of students were to be based entirely on ability, and no consideration of a social or political nature, which must necessarily enter into any appointment at collegiate institutions, was to be made. The founders of the institute made this clear in a letter addressed to the trustees, as follows:

"It is our hope that the staff of the institution will consist exclusively of men and women of the highest standing in their respective fields of learning, attracted to the institution through its appeal as an opportunity for the serious pursuit of advanced study and because of the detachment it is hoped to secure from outside distraction.

"It is fundamental in our purpose, and our express desire, that in the appointments to the staff and faculty, as well as in the admission of workers and students, no account shall be taken, directly or indirectly, of race, religion, or sex. We feel strongly that the spirit characteristic of America at its noblest, above all, the pursuit of higher learning, cannot admit of any conditions as to personnel other than those designed to promote the objects for which this institution is established, and particularly with no regard whatever to accidents of race, creed, or sex."

It was also intended to free the faculty of this institute as far as possible from all administrative and pedagogical duties, so that they could concentrate on their academic work. In the letter the founders also said:

"It is our desire that those who are assembled in the faculty of the institution may enjoy the most favorable opportunities for continuing research in their particular field and that the utmost liberty of action shall be afforded to the said faculty to this end."

In his address at the organizing meeting Flexner emphasized particularly that the members of the institute were to have better living conditions than in most universities. He said:

"The sacrifices required of an American professor and his family are to a high degree deterrent. The conditions provided are rarely favorable to severe prolonged and fundamental thinking. Poor salaries frighten off the able and more vigorous and compel the university instructor to eke out his inadequate income by writing unnecessary textbooks or engaging in other forms of hackwork. . . . It is therefore of utmost importance that we should set a new standard."

It thus became the policy of the institute to have a faculty consisting of a few excellent but well-paid members.

At first no decision was made as to which subjects would be cultivated at the institute, but if the principles laid down by the founders and Dr. Flexner were to be realized, the limited means that were available made it necessary that the institute restrict its activities at first to a certain special field. After a good deal of reflection and consultation Flexner decided to devote the institute first to mathematical sciences. He was led to this choice by three reasons. Firstly, mathematics is fundamental; secondly, it requires the least investment in equipment and books; and thirdly, it became obvious to Flexner that he could secure greater agreement upon those who were considered the outstanding leaders in the field of mathematics than in any other field.

Until the institute could have its own building, President Hibben of Princeton University turned over to Flexner a part of Fine Hall, the mathematics building on the Princeton campus. The beautiful campus with the shady trees and the buildings in the Gothic style of the English universities presented a stimulating environment. Also, the institute obtained a certain point of departure for its activities by collaborating with the mathematicians of the university. It was expected that as time went on, outstanding men from the entire world who had already obtained the doctorate in mathematics would come to Fine Hall.

From the very beginning it had been the idea of the founders that it should somehow have a cloistered character. As Flexner once expressed it: "It should be a haven where scholars and scientists may regard the world and its phenomena as their laboratory without being carried off in the maelstrom of the immediate." This seclusion of the institute was increased in 1940 when it moved away from Fine Hall and the Princeton campus to its own building, situated a few miles outside of the town of Princeton.

2. *Einstein's Decision to Join the Institute*

Flexner first set out to look for the great masters who were to form the basis of his institute. He traveled through America and Europe looking for men of such rank who were available. In the course of these journeys he came to Pasadena in the winter of 1932. There he discussed the matter with R. A. Millikan, the famous physicist, who said to him: "You know that Einstein is a guest here at present. Why don't you tell him about your plan and hear his opinion?" At first Flexner was rather hesitant about discussing such questions concerning teaching and administration with a man who had already become a legend. He was afraid to approach Einstein because he was "a too much lionized man." Millikan told him, however, that Einstein was a man who was interested in all projects for improving the training of young scholars and who liked everything that was new and bold. "I will tell him about you immediately. Look him up at the Athenaeum." This is the faculty club of the California Institute of Technology, situated in the midst of a beautiful palm garden, where foreign scholars stayed as guests.

Flexner described this visit as follows:

"I drove over to the Athenaeum where he and Mrs. Einstein were staying and met him for the first time. I was fascinated by his noble bearing, his simply charming manner and his genuine humility. We walked up and down the corridor of the Athenaeum for upwards of an hour, I explaining, he questioning. Shortly after twelve, Mrs. Einstein appeared to remind him that he had a luncheon engagement. 'Very well,' he said in his kindly way, 'we have time for that. Let us talk a little longer.' "

At this time Flexner did not yet consider Einstein himself for this institute. He wanted only to hear his opinion about the plan. They agreed to meet again early the next summer at Oxford, where Einstein was to lecture.

As they had planned, Einstein actually met Flexner on the beautiful lawn of the quadrangle of Christ Church College at Oxford, where Einstein was staying. Flexner describes the meeting:

"It happened to be a superbly beautiful day and we walked up and down, coming to closer and closer grips with the problem. As it dawned on me during our conversation that perhaps he might be interested in attaching himself to an institute of the proposed kind, before we parted I said to him: 'Professor Einstein, I would not presume to offer you a post in this new Institute, but if on reflection you decide that it would afford you the opportunity which you value you would be welcome on your own terms.' "

They agreed that during the summer Flexner would come to Berlin to continue the talks. It was the summer of Papen's interim government in Germany, the summer when the German Republic was already dead and led only a ghostly existence. Einstein saw the future with complete clarity and had decided to keep the road to America open for himself.

When Flexner came to Berlin, Einstein was already living in his country home at Caputh near Potsdam. It was the same summer and the same house of which I have already spoken in Chapter X. Flexner arrived at Einstein's country house on a Saturday at three in the afternoon. He describes his visit as follows:

"It was a cold day. I was still wearing my winter clothes and heavy overcoat. Arriving at Einstein's country home, beautiful and commodious, I found him seated on the veranda wearing summer flannels. He asked me to sit down. I asked whether I might wear my overcoat. 'Oh yes,' he replied. 'Aren't you chilly?' I asked, surveying his costume. 'No,' he replied, 'my dress is according to the season, not according to the weather, it is summer.'

"We sat then on the veranda and talked until evening, when Einstein invited me to stay to supper. After supper we talked until almost eleven. By that time it was perfectly clear that Einstein and his wife were prepared to come to America. I told him to name his own terms and he promised me to write to me within a few days."

As was his custom, Einstein, wearing a sweater and no hat, accompanied his visitor through the rain to the bus station. The last thing he said on bidding Flexner farewell was: "I am full of enthusiasm about it."

Einstein soon communicated the conditions under which he would take the new position in a letter to Flexner, who found them much too modest for such an institute and for a man like Einstein. He requested that the negotiations be left to himself and to Mrs. Einstein. The contract was concluded at this time. Einstein pointed out that he was obliged to spend the winter of 1932–3 in Pasadena and could not go to Princeton until the fall of 1933. At that time he still had intentions of spending a part of every year in Berlin, for he preferred not to be unfaithful to his friends in the world of physics there. But he was very much aware of coming events. When the Nazi revolution took place early in 1933, the way had already been prepared for his emigration to America, and in the winter of 1933 Einstein entered upon his new position at the Institute for Advanced Study, which Flexner had founded at Princeton. Now there was naturally no further mention of spending a part of the year in Berlin. Einstein moved to Princeton to become a permanent resident and citizen of the United States. There were still, however, a number of stages through which he had to pass in order to achieve this goal. He had entered the country only as a visitor and for the present had no legal right to remain here permanently, to say nothing of becoming a citizen.

---

3. *Einstein's Activities at the Institute*

The institute that Einstein joined was in some respects similar to the Kaiser Wilhelm Institute to which he had belonged in Berlin. Thus he once again occupied a position that in a certain respect had always appeared distasteful to him. As I have already mentioned, he always regarded it as an uncomfortable situation for anyone to be paid only for his research work. One does not always have really valuable ideas, so that there is a temptation to publish papers of no special value. Thus the scientist is subject to a painful coercion. But when one is a teacher with a moderate load one has every day the consoling feeling of having done a job that is useful to the society. In such a situation it is satisfying to carry on research for one's own pleasure during the leisure hours, without compulsion.

On the other hand, however, a man of creative ideas like Einstein chafed under the daily routine of teaching. He found the idea of teaching very noble, but when he was actually offered a position where he would be able to devote himself entirely to research, he was unable to refuse it. At the new Institute he was able to guide talented students who had already acquired the doctorate in carrying on their researches. In consequence, however, his contact with students was restricted to a very small group. Einstein often vacillated between a feeling of satisfaction at being spared any routine work and a certain feeling of loneliness due to his isolation from the great mass of students.

This divided attitude was quite in accord with his divided attitude toward contact with his fellow men in general, which has already been mentioned repeatedly. This division, which has played a large part in his entire life, manifested itself also in his attitude to his environment in Princeton. It would have been simple enough for him to give lectures or to organize a seminar, which many of the students might have attended. Einstein, however, felt that it would not be fair for him, a man with an international reputation, to enter into a contest with the professors of the university, some of whom were still quite young. They could with some justification regard it as "unfair competition." At any rate Einstein very punctiliously avoided any such competition for students. It is possible, however, that he exaggerated in his mind the touchiness and ambition of his colleagues, since many would gladly have taken advantage of the presence of such an outstanding scientist in Princeton to learn from him themselves. As things were, his presence there has not been utilized so much as it might have been. No one, perhaps not even Einstein himself, can say to what degree this situation is due to his consideration for others and his aversion toward more intimate contact with people.

By and large, at Princeton Einstein took up his researches where he had left off in Berlin; this is true both of the problems themselves and of the way in which he dealt with them. It was always very characteristic of him to be independent of his environment. Just as at the time of our meeting in Berlin twenty-five years before it had been a matter of indifference to him whether he was working at his problems in his study or on a bridge in Potsdam, so now it made no difference to him that he had moved his office from the western suburb of Berlin to the distinguished American university town of Princeton beyond the ocean.

Three groups of problems occupied Einstein during this period. In the first place there was the desire to develop his special and general relativity theories of 1905, 1912, and 1916 into an ever more logically connected structure. In one important point Einstein succeeded in making a great advance at Princeton. It will be recalled that Einstein regarded the gravitational field as a geometrical property of space, which can be called in a word the "curvature." This curvature is determined by the presence of matter in space and can be computed from the distribution of matter. If the curvature of space, or, in other words, the gravitational field, is known, then one also knows how a body that is present in this space will move. This is given by the "equations of motion," which can be stated briefly as follows: A body moves in such a manner that the representation of the path in a four-dimensional space-time continuum is a geodesic (shortest) line. This would be entirely satisfactory if one assumed that matter and force field were two completely different entities. But one is driven closer and closer to the conception that the mass of a particle is actually nothing but a field of force that is very strong at this point. Consequently "motion of mass" is nothing but a change of the force field in space. The laws for this change are the "field equations" — that is, the laws that determine the force field. But if the motion of the body is already determined by the field equations, then there is no more room for special laws of motion. One cannot make a supplementary assumption, in addition to the field equations, that masses move along geodesic lines. Instead these equations of motion must already be contained in the field equations.

C. Lanczos, Einstein's co-worker in Berlin, had sketched the idea of deriving the laws of motion mathematically from the field equations. His derivation, however, did not appear satisfactory to Einstein, and in Princeton he succeeded in showing in a completely convincing way that only the field laws need be known in order to be able to derive the laws of motion from them. This is regarded as a confirmation of the idea that matter is nothing but a concentration of the field at certain points.

I have already mentioned that Einstein liked to have the assistance of young physicists or mathematicians, especially when he dealt with involved mathematical computations. From Berlin he had brought with him the Viennese mathematician Walter Mayer, who soon obtained an independent position at the Institute of Advanced Study and no longer collaborated with Einstein. During the first few years of Einstein's residence a very talented Polish physicist, Leopold Infeld, came to Princeton, where he remained several years and with whom Einstein worked out the aforementioned proof of the "unity of field and matter."

Einstein liked to discuss with Infeld all kinds of problems, including the fundamental problems of physics and their development. These conversations gave rise to the book by Einstein and Infeld, *Evolution of Physics,* which has achieved a wide circulation. It is certainly one of the best presentations of the fundamental ideas of physics for the public at large.

Infeld wrote also an autobiography entitled *Quest: The Making of a Scientist.* This book contains much about Einstein's life at Princeton as seen by a keen observer and competent collaborator.

A second group of problems with which Einstein was intensely occupied at this time is the criticism of the development of the quantum theory, which has been described in Ch. IX. Einstein felt impelled to show by concrete examples that the quantum theory, in the "Copenhagen" form in which it had been formulated by Niels Bohr, did not describe a "physical reality," such as a field, but only the interaction of the field with a measuring instrument. A paper that Einstein published together with N. Rosen and B. Podolsky, two young physicists, was particularly important in this discussion. This paper shows by a simple example that the way in which the quantum theory describes the physical condition in a certain spatial area cannot be called a complete description of physical reality in this area.

This work stimulated Niels Bohr to formulate more clearly than he had previously done his standpoint on the question of physical reality. Bohr now rejected definitely all the "mystical" interpretations to which his theory had been subjected. Among these was the conception that the "real state" in a spatial area is "destroyed" by observation, and similar ideas. He now stated clearly that the quantum theory did not describe any property of a field, but an interaction between field and measuring instrument. It is plain that one could not decide between the conceptions of Einstein and Bohr by general logical considerations, since they were not opposed assertions, but rather opposed proposals.

Einstein proposed to retain tentatively a kind of description of the physical state in an area of space which was not too far from the way in which the language of daily life describes reality. This means that he proposed to describe the physical state in an area in such a manner that the description itself need not state with which measuring instrument it was obtained. Einstein has been well aware that it would not be absurd to abandon this kind of description where the laws of physics are formulated in terms of a "field"; but he would abandon it only if it became necessary beyond any doubt.

The third and most exciting problem was his attempt to find the actual physical field that permits a formulation of physical laws for the subatomic phenomena in a form that is a generalization of the equations of the electromagnetic and gravitational fields. Einstein has collaborated in this task with two young men, one called Bergmann, the other Bargmann, a similarity that gave rise to many jokes.

Every forenoon Einstein went regularly to his office at the Institute for Advanced Study, where he met either Peter Bergmann or Valentin Bargmann or both of them. Einstein suggested to them various ways of conceiving the structure of space, not only as four-dimensional, but sometimes also as five-dimensional, so that the magnitudes that describe this geometrical structure could also furnish a description of the unified physical field of force. The real force field would then be found, if one could find relations between the described magnitudes, from which the actual laws of observable phenomena can be derived in all domains of physics, including atomic and nuclear physics.

The difficulties of this task proved to be even greater than had been supposed. At present it seems that all the paths that have previously been tried do not lead to the goal. Recently, Einstein has probed new field equations and he has by no means abandoned the hope that electrons and protons will turn out to be just particular fields. Despite the tremendous range of experimental confirmation of Bohr's "positivistic" Theory it remains, according to Einstein, still an open question, whether it is not possible to derive the same observable facts from a field theory and to save the historic conception of a physical reality independent of the devices of observation and measurement.

Besides the regular work connected with the Institute, Einstein had to occupy a part of his time as an adviser of young men with interest or ambition in science. It had often been Einstein's fate to be judged not only as an individual, but as a type — indeed, even more, as the symbol of a certain group of people. This fate was all the more painful for him as there was nothing that he liked less than to be classified as a member of a party or a group. As he had come out courageously for the cause of the Jewish people he has been expected to play the part of a leader or at least of a representative of his people both by the enemies of the Jews and by the Jews themselves. The life of Einstein has been regarded as symbolizing the fate of a people, often talented, but often attacked and driven into isolation. Therefore among the people who looked for Einstein's advice there were many young Jews who wrote letters to him appealing for his help. In some degree he has played the role among the Jews that Tolstoy at one time had played among Russian youth. Poor young Jews looked upon Einstein as one of their people who had made good and who was so world-renowned that boundless power and wealth were ascribed to him. This was, I dare say, a great mistake. Neither his fortune nor his influence has corresponded even remotely to his fame.

Very often young people of any background turned to him for advice about beginning an academic career for which they felt they were equipped instead of turning to some mechanical work in an office or a shop. Einstein was always ready to advise what he considered proper and was interested in each one's personal situation. However, as we have learnt, Einstein also believed that it was a good thing to earn a living by means of a "cobbler's trade" and to devote one's spare time to study.

Einstein never liked to speak about the material and moral help he provided for distressed people. I recall several cases, however, that I was able to observe myself. Einstein remained interested in students whom he had helped to enter a university, and continued to watch them as they progressed with their studies. He advised them with which teachers to study, which books to read, and even sent them books himself. I remember one such case very clearly.

It concerned a student from one of the Balkan countries. Upon Einstein's advice he had applied to the university in Prague, where he was admitted. Einstein asked me to take an interest in him and so he consulted me when he had trouble. The student lived on a stipend that he received from a big manufacturer in his native land. But this money, which barely sufficed for himself, the student used to enable his brothers and sisters to study as well. The fact that one of the greatest men of our time was watching his studies was the great event of his life and filled even the minutest experience with a remarkable splendor. When the young man first turned to Einstein, the latter was still in Berlin, but when the young man arrived in Prague Einstein was already in America. The student wrote to Einstein telling him about every phase, even the most trivial, of his studies; and frequently he received answers from America that gave him extremely detailed advice. When the student met with difficulties in his relations with teachers or fellow students, he asked Einstein's advice as to how he should behave. Einstein usually advised him to be conciliatory. This was certainly very good advice for this young man, as he became involved in various conflicts in the unfamiliar environment. He was naturally filled with pride because he was distinguished from all the other students of physics by the fact that he corresponded personally with the greatest physicist of our time.

There is little wonder that occasionally a student in this situation imagined himself as Einstein's representative to such a degree that he regarded all insults to him as insults to Einstein. He would even feel that he was a martyr, happy at being permitted to suffer for Einstein, and finally even came to believe that by being connected with Einstein he was making a sacrifice and getting himself into trouble.

4. *Refugee Scholars*

As the persecution of the Jews increased in Germany and her satellite countries, the number of scientists, writers, artists, teachers, and others who wished to find haven in the United States grew larger and larger. As when a large quantity of good merchandise is thrown on the market at reduced prices, economic repercussions occur, even inflation, so when these refugee scholars offered themselves, great difficulties were encountered.

The new immigration began while the United States was still in the midst of the great economic crisis. This was, of course, not an accidental coincidence since without the worldwide depression the Nazi revolution in Germany would not have occurred. As the number of immigrants increased, fantastic rumors began to be spread about them. It was frequently said that the refugees were not pioneers; they did not perform any constructive work as the earlier immigrants had done, but they wanted only to get rich without working or to live on charity. Many regarded and feared them as professional competitors, many simply used them as scapegoats whom they could blame for various ills. Skillful agitators were even able to convince people that enormous numbers of such immigrants would soon change the national and racial composition of the people of the United States.

When Bertrand Russell, the English mathematician and philosopher, because of his critical attitude to traditional views on marriage and religion, was prevented from being appointed as professor of philosophy at the College of the City of New York, Einstein backed him. He felt that it was harmful for the development of science when attacks of personal and political opponents could prevent the appointment of a scientifically outstanding professor. Russell's enemies, however, used Einstein's support for their own aims. They wrote letters to newspapers containing such statements as: "How dare the 'nudist' Russell and the 'refugee' Einstein interfere in the family life of the United States!" The use of the words "nudist" and "refugee" as equally disparaging characteristics is noteworthy.

Every institution that wanted to appoint one of the refugee scholars was in a dilemma. On the one hand, the American universities were quite ready to help the victims of political persecutions and were glad to have the opportunity of acquiring men of great ability, but, on the other hand, they had a responsibility to their own graduates who were looking for academic positions. It would have been a severe disappointment for them to have positions unexpectedly filled by scholars from Europe who were naturally older and had greater reputation.

This situation also placed those refugee scholars who had already obtained a position in a difficulty. They felt morally obligated to help their countrymen and fellow sufferers who had been less fortunate, but they also felt obliged to look after the interests of their students primarily. Some of them even went so far as to say that it was the duty of every refugee scholar who had a position to see to it that no other refugee obtained one at the same institution.

For Einstein the situation was even more difficult. Here again he came to be regarded as the symbol and leader of the entire group of refugee scholars. The friends of the refugees upheld Einstein as an example of the outstanding men who were coming to the United States, while their opponents felt themselves compelled to disparage him in order to oppose the refugee group. The refugees themselves looked upon Einstein as their natural leader. They felt that with his fame he would somehow be able to help them, and they turned to him for help.

Einstein received hundreds of letters from scholars in Europe who wanted to emigrate and who asked his aid in getting them a position or an "affidavit of support," required under the American immigration regulations. Einstein tried hard to help them, and he even made out such affidavits himself for many. Others immediately turned to Einstein on their arrival in America. He did the best he could, but naturally the number of persons whom he was actually able to help was very small in comparison with the enormous number who appealed to him.

In recommending foreign scholars for positions, Einstein as always had only two considerations in mind: the immediate feeling of sympathy for every suffering person, and the conviction that the pursuit of science should be assisted wherever possible. He was always ready to write recommendations for these people. He thought that if a foreign scientist was needed, his recommendation would be of some help, and if this were not the case, it would not hurt either the person recommended or the institution.

Einstein might have done more for the refugees if he had undertaken to study the situation at various universities and to take advantage of the personal, economic, and political factors involved, but such an action was not possible for him. The people who are the most outstanding intellectually and also the kindest are not always very practical. This explains the contradictory opinions about Einstein. Some people felt that he was kind and devoted, others that he cared little for the fate of individuals.

While co-operating sincerely in charitable social and political organization Einstein will suddenly tell you: "Sincerely speaking I have never been much interested in people but only in things." And if you ask him what he meant by "things" he would say: "physical phenomena and methods to handle them."

The psychological situation of these new refugees also had its difficulties. Many came from Germany, which they had always considered their native land and with whose intellectual and cultural life they felt themselves united. They had been driven out, but that did not mean that they had therefore lost all connection with it. They came to a foreign country that gave them a friendly reception and made it possible for them to start a new life, which was sometimes even better than the life in their former country. If they laid too great an emphasis on their connection with German culture, they could easily arouse a feeling of antagonism to themselves in the new country.

On the other hand, owing to the circumstances leading to their emigration, they were strongly opposed, both politically and culturally, to the ruling circles in Germany. As a result, they were accused on the one hand of propaganda in favor of German culture, and on the other hand of carrying on hate propaganda that might create enmity between the United States and Germany or even involve them in war. Remarkably enough, these contradictory accusations were often made at the same time.

Einstein himself was often surprised that the new immigrants from Germany still remained so much attached to their old country. It was a special puzzle to him why the Jewish refugees, who had suffered so much in Germany, still had such a strong yearning for that country. As Erika and Klaus Mann reported, Einstein on one occasion told this story:

"I met a young German lawyer who is living in New York, a so-called Aryan, and asked him whether he was homesick. 'Homesick?' he said. 'I? What for? I am not a Jew.'

"Isn't that a good one?" Einstein added. "Isn't that typical? Isn't the nationalism of the Jews sentimental and lachrymose, a sullen and morose love for a country such as is to be found only among people who do not feel sure which country is theirs?"

"I am also a Jew," continued Einstein, "but yet everything seems to me so fine in America that I am not homesick for any country, to say nothing of Herr Hitler's Germany."

We know Einstein's aversion against the inhuman mechanical attitude of the German ruling caste under the Kaiser, let alone under Hitler. Equally strong, however, is his love for the German music of Bach and Mozart. In certain respects, perhaps, he even shares the tastes of the German nationalists in art. He dislikes "modern" music and finds it rather repugnant. Generally, he likes everything German that derives from the spirit of the pre-Bismarckian and pre-Wilhelminian period. He has been happy with visitors imbued with the spirit of classical German music and literature. He is even quite sympathetic to the Kantian philosophy, partly perhaps because of its emotional relationship with that period of the German spirit. He has this sympathetic feeling for it although on purely scientific grounds he has rejected it in all essential points.

I have been struck by the fact that despite his emphasized hostility to the spirit of a Germany ruled by Prussian militarists, he has always been fond of conversing with men — for instance, German-American ministers, in whom the older German spirit had somehow been preserved.

In America Einstein had often been regarded officially as a leader of the Jewish people. When the World's Fair was opened in New York in 1939–40, Palestine was represented by a pavilion. Since it was customary, at the opening of a pavilion, for the ambassador of the particular country to deliver an address, the question arose who should deliver such an address at the opening of the Palestinian pavilion. The choice did not fall upon a political leader of the Zionists, nor on a rabbi, but instead on Einstein, who was thus officially recognized as a kind of spiritual leader of the Jews.

---

5. *Einstein's Attitude toward Religion*

To understand Einstein in its attitude to the Jewish people, one must understand his attitude to traditional Biblical religion and to religion in general. Would not a man like Einstein, whose unsparing criticism had removed the last remnants of medieval semi-theological conceptions from physics, assume a purely critical attitude to the religion of the Bible? Ever since his arrival in America this aspect of his personality has been much in the limelight. In this country people are more interested than in Europe in the problem of the relation between science and religion, and they feel more strongly the need for a mutual understanding between them.

Einstein's attitude toward traditional religion is related in turn to his divided attitude to social relations in general. When I first became acquainted with Einstein, around 1910, I had the impression that he was not sympathetic to any kind of traditional religion. At the time of his appointment to Prague he had again joined the Jewish religious community, but he looked at this act rather as at a formality. At this time, too, his children were about to enter elementary school, where they would receive religious instruction. This was a rather difficult problem as he belonged to the Jewish religion and his wife to Greek Orthodox. "Anyway," said Einstein, "I dislike very much that my children should be taught something that is contrary to all scientific thinking." And he recalled jokingly the manner in which school children are told about God. "Eventually the children believe that God is some kind of gaseous vertebrate." This was an allusion to a saying of the German scientist and philosopher Ernst Haeckel that was then current.

At that time a superficial observer would easily have settled the question of Einstein's attitude to religion with the word "sceptical." Perhaps characteristic of this attitude is a remark that Einstein made to an orthodox Jew whom he once met at a police station in Prague, where I had gone with him to obtain a visa for a passport. The man asked Einstein if he knew a restaurant in Prague where the food was strictly kosher — that is, prepared according to the ritualistic precepts. Einstein mentioned the name of a hotel that was known to be kosher. The man asked Einstein: "Is this hotel really strictly kosher?" This annoyed Einstein somewhat and he said seriously: "Actually only an ox eats strictly kosher." The pious man was hurt and looked indignantly at Einstein. The latter, however, explained that his statement was not offensive at all but quite objective and innocent: "An ox eats grass, and that is the only strictly kosher food because nothing has been done to it."

Einstein's attitude reflects often the immediate response of a genius which is similar to that of an intelligent child. The world is not judged in the traditional way, but as reason suggests. If this judgment is expressed without any of the traditional euphemisms, it is often called "cynical"; but it should be called rather "sincere with a sense of humor."

Einstein was once told that a physicist, whose intellectual capacities were rather mediocre, had been run over by a bus and killed. He remarked sympathetically: "Too bad about his body."

On another occasion Einstein was invited by a committee organized to honor a well-known scholar to take part in the celebration of his seventieth birthday and to address the gathering. Einstein replied to this committee: "I hold the man whom you are honoring in high esteem and I like him very much. For this reason I will arrange a dinner in his honor all alone at my home on his birthday. Since no audience will be present, I will simply keep the speech to myself. Wouldn't it be more convenient for you and the scholar whom you are honoring if you did the same?"

His manner of speech is often an expression of the urge to make the serious things in the world tolerable by means of a playful disguise, a form of behavior that is ultimately the basis of all artistic activity. The use of such caustic words was for

281

Einstein an artistic way of coming to terms with the world, like the playing of a Mozart sonata, which also represents the evil of the world in a playful manner. In a certain sense all of Mozart's music might be called "cynical." It does not take our tragic world very seriously and reflects it in gay, youthful rhythms.

To understand Einstein's views on religion seriously it is good to start from his conception of physical science and of science in general. As I have already repeatedly emphasized, the general laws of science, according to Einstein, are not products of induction or generalization but rather products of free imagination which have to be tested by physical observations. In his Oxford address Einstein asks:

"If it is the case that the axiomatic basis of theoretical physics cannot be an inference from experience, but must be free invention, have we any right to hope that we shall find the correct way? Still more — does this correct approach exist at all save in our imagination?"

To Einstein the physical theory is a product of human inventiveness, the correctness of which can be judged only on the basis of its logical simplicity and the agreement of its observable consequences with experience. This is exactly the description of a theory and the criterion of its validity which has been advocated by the *Logical Positivists*. To them the belief in the "existence of a correct theory" means the "hope to make a certain invention." The expression "the correct form of a theory" has no more meaning than "the correct form of an airplane" what is obviously a meaningless expression.

But here Einstein deviates definitely from the conception of Logical Positivism. In his Oxford lecture he replied to the question whether there is a "correct way" as follows:

"To this I answer with complete assurance that in my opinion there is the correct path and, moreover, that it is in our power to find it. Our experience up to date justifies us in feeling sure that in nature is actualized the idea of mathematical simplicity.

"It is my conviction that pure mathematical construction enables us to discover the concepts and the laws connecting them, which give us the key to the understanding of the phenomena of nature. Experience can, of course, guide us in our choice of serviceable mathematical concepts; it cannot possibly be the source from which they are derived.

"In a certain sense, therefore, I hold it to be true that pure thought is competent to comprehend the real as the ancients dreamed."

Here Einstein even goes so far as to use the language of idealistic philosophy, of the advocates of an aprioristic knowledge —

that is, knowledge independent of experience — although he has been a decided opponent of this philosophy. Nevertheless, in order to emphasize as strongly as possible his opposition to some oversimplifications current under the name of "positivism" he employs a mode of expression that can easily be misunderstood by those who have only a superficial knowledge of Einstein's views.

The difference between Einstein's views and those "dreams of the ancients" to whom he felt related is the following: According to the views of the ancient philosophers the power of intuition suffices to advance propositions that do not need to be tested by experience. But this is not what Einstein actually means. He means that the inventive faculty presents us with various possibilities for the construction of mathematical theories, among which only experience can decide.

The conviction of which Einstein spoke, and for which, naturally, no cogent reasons can be given, is the following: among the theories there will some day be one which in its logical simplicity as well as in its simple representation of observation will be so greatly superior to all rival theories that everyone will recognize it as the best in every respect. This conviction is nothing but an expression of scientific optimism. It is an expression of belief in a certain constitution of observable nature, which has been often called a "belief in the rationality of nature."

The existence of such logical pictures of nature is a characteristic that is not self-evident, but which we recognize by experience and which we may call the "rationality of nature," if we prefer to employ the terminology of traditional philosophy. This terminology is usually employed when one wishes to express one's sympathy with certain feelings that are customarily expressed with great beauty in the language of that philosophy. Amazement at this *rational aspect of nature* turns into admiration; and this admiration is, in Einstein's opinion, one of the strongest roots of religious feeling.

When we speak of the existence of a logical system that corresponds to natural processes, the term "existence" means in everyday language only that there are thinking beings similar to men which can imagine such a system. If we speak of the "existence" of such a system without relating it to a thinking being, it is an obscure mode of expression. If we do connect it with a thinking being, we imagine more or less vaguely a being similar to man with superior intellectual capacities. Consequently, to speak of the "rationality" of the world always means to think vaguely of a spirit superior to man and yet similar to him. In this way Einstein's conception of nature is related to what is usually called a "religious" conception of the world.

Einstein knows very well that this is not a statement about nature that is in any way scientific, but that it expresses a feeling aroused by the contemplation of nature. In this connection he once said:

"The most beautiful emotion we can experience is the mystical. It is the sower of all true art and science. He to whom this emotion is a stranger, who can no longer wonder and stand rapt in awe, is as good as dead. To know that what is impenetrable to us really exists, manifesting itself as the highest wisdom and the most radiant beauty, which our dull faculties can comprehend only in their most primitive forms — this knowledge, this feeling, is at the center of true religiousness. In this sense, and in this sense only, I belong to the ranks of devoutly religious men."

According to Einstein's conception, it is particularly the scientist in the field of natural science, and especially in the field of mathematical physics, who has this mystical experience. Here is the root of what Einstein calls "cosmic religion." He once said:

"The cosmic religious experience is the strongest and the noblest, deriving from behind scientific research. No one who does not appreciate the terrific exertions, the devotion, without which pioneer creation in scientific thought cannot come into being can judge the strength of the feeling out of which alone such work, turned away as it is from immediate practical life, can grow.

"What deep faith in the rationality of the structure of the world, what a longing to understand even a small glimpse of the reason revealed in the world, there must have been in Kepler and Newton!"

In recent years the view has frequently been put forth that the physical theories of the twentieth century, especially Einstein's theory of relativity and the quantum structure of energy, have a great significance for the mitigation of the conflict between religion and science. Since Einstein has spoken of a "cosmic religion" based on science he has been often quoted as an advocate of that view. This, however, is a great misunderstanding. With his clear insight into the logical structure of a scientific theory, he has never encouraged the religious interpretation of recent physics which became current by the popular books of such scientists, as Jeans and Eddington.

For Einstein religion is both a mystical feeling toward the laws of the universe and a feeling of moral obligation toward his fellow men as well. Nevertheless, the strictly logical-empirical character of his thought prevents him from assuming a scientific or apparently scientific link between these two feelings. We may feel a hint of it in music, which expresses what cannot be formulated in words.

This feeling, however, has been misunderstood by some people, since Einstein has never placed any importance on the formal aspects of religion. It was striking how readily Einstein used the word "God" as a figurative expression, even in physics. It will be recalled that he had repeatedly expressed his rejection of the statistical conception of physics with the statement: "I cannot believe that God plays dice with the world." It is certain that the word "God" is used here only as a figure of speech and not in a theological sense. Other physicists, however, do not employ this figure of speech with equal readiness. One of Einstein's finest remarks, which is recorded on a wall in the Institute of Advanced Study at Princeton, expresses his conception of the nature of physical science by means of the same figure of speech. Einstein wants to say that from a mathematical standpoint the system of physical laws is very complex, and that to understand it very great mathematical capacities are required. Nevertheless, he has hope that nature actually obeys a system of mathematical laws, and that the human mind can find these laws if it allows itself to be led by its scientific judgment. All this is expressed by the aforementioned remark:

"God is sophisticated, but he is not malicious."

In the fall of 1940 a conference was held in New York to discuss what contributions science, philosophy, and religion could make to the cause of American democracy. Einstein was among those asked to address the conference. At first he did not want to write anything, as he disliked to attract public attention to himself, especially in political matters. Nevertheless, as the aim of the conference appealed to him, he allowed himself, even though he would not appear and speak in person, to be induced to send a written contribution, entitled "Science and Religion." In it he said:

"The main source of the present-day conflicts between the spheres of religion and of science lies in the concept of a personal God. It is the aim of science to establish general rules which determine the reciprocal conceptions of objects in time and space. . . . It is mainly a program, and faith in the possibility of its accomplishment in principle is only founded on partial success. But hardly anyone could be found who would deny these partial successes and ascribe them to human self-deception. . . .

"The more a man is imbued with the ordered regularity of all events, the firmer becomes his conviction that there is no room left by the side of this ordered regularity for causes of a different nature. For him neither the rule of human nor the rule of divine will exists as an independent cause of natural events. To be sure the doctrine of a personal God interfering with natural events could never be refuted, in the real sense, by science, for this doctrine can always take refuge in those domains in which scientific knowledge has not yet been established. . . .

"To the sphere of religion belongs the faith that the regulations valid for the world of existence are rational, that it is comprehensible to reason. I cannot conceive of a genuine scientist without that profound faith. The situation may be expressed by an image: science without religion is lame, religion without science is blind."

There is apparently nothing sensational or shocking in these sentences. Those scientists who were willing to concede to religion an important place in human life have generally found that Einstein formulated just what they thought. On the other hand, there are certainly many scientists who take it much amiss that Einstein even mentions religion and spirituality in the same breath with science.

Suddenly, however, a number of people appeared with the cry: "Einstein wants to deprive us of our personal God." "It is this very personal element in God," they said, "that is most precious to man." Einstein received innumerable letters, many containing vehement accusations to the effect that he wanted to rob people of such a beneficial faith. Letters to the editor appeared in newspapers, in which the writers protested against permitting a "refugee" to meddle with the belief in God.

There were Christian clergymen who asserted that the expression "personal God" was characteristic of the Christian God in contradistinction to the Jewish God, and that Einstein was carrying on a polemic against the Christian conception of God. Actually Einstein knew nothing of these subtleties of Christian and Jewish theology. On the contrary he wanted to emphasize the common ground of liberal Judaism and liberal Christianity in their conception of God. But here again, as in so many instances, his well-meant intentions involved him in odious and malicious polemics, which he could not have foreseen.

As in most other points, Einstein advocated practically positivistic views concerning the relations of the exact sciences and of science in general to human conduct. To the question whether the goal of human life could be derived from science only, Einstein, like positivism, replied with a decided "No." Like logical positivism, Einstein is of the opinion that no matter what degree of mathematical simplicity and beauty the laws of nature exhibit, no matter how well they reflect observation, they can never tell us what man's aims in life should be. From natural laws we learn only how nature behaves, how we can utilize these forces to realize human aims, not what these aims should be.

These aims man can learn only by example and indoctrination. It is this indoctrination that Einstein believes to be the task of the church, not the preaching of a certain conception of nature.

Because Einstein is profoundly convinced that science, even when most highly developed, cannot present man with a goal, he is far from disputing the usefulness of church organizations. He does not care for religious ritual, but he realizes the value of churches and religious services as means of education; and in so far as the ritual increases the effect of indoctrination, he has learned to appreciate the value of religious ceremony.

Einstein's views on the responsibility of the church for moral education may be seen, perhaps, in an address that he delivered in the summer of 1939 in the theological seminar at Princeton before an audience of clergymen and students of theology. The title of the lecture was "The Goal." Among other things he said:

"It is certainly true that principles cannot be more securely founded than on experience and consciously clear thinking. In this one will have to agree absolutely with the extreme rationalists. The weak point of the conception, however, lies in the fact that those principles which are decisive and necessary for our actions and judgments of value cannot be obtained only in this scientific way. The scientific method cannot teach us anything but the conceptual comprehension of the reciprocal relations among facts. The endeavor to obtain such objective knowledge is one of the loftiest aspirations of which man is capable, and you will certainly not suspect me of underrating the heroic efforts and achievements of the human mind in this field. On the other hand, however, it is clear that no path leads from a knowledge of that which is to that which should be. No matter how clear and perfect our knowledge of present existence, no goal for our human aspirations can be inferred from it. . . . No matter how splendid the knowledge of truth as such may be, as a guide it is so impotent that it is not even able to establish the justification and the value of this very striving for a knowledge of the truth. . . .

"Reason apprises us of the interdependence of aims and values. What thought alone cannot give us are the ultimate and most fundamental goals, in terms of which the more secondary ones are oriented. The setting up of the most fundamental goals and valuations and their establishment in the life of the individual seem to me to be the most important function of religion in the social life of man. If one should ask whence these fundamental goals receive their authority, since they are not set up by reason and cannot be founded upon it, one can only answer that they do not come into existence as a result of argument and proof, but instead by revelation, and through the actions of strong personalities. One should not attempt to prove them, but rather to recognize their essence as clearly and purely as possible.

"The most fundamental principles of our aspirations and valuations are given to us in the Judeo-Christian religious tradition. It is a lofty goal. . . . When one divests this goal of its religious form and regards only this purely human side, it may be expressed as follows:

"Free and self-responsible development of the individual so that he will freely and joyfully put his energies at the service of the community of man. If attention is paid to the content and not to the form, the same words may be considered as the expression of the fundamental democratic principle. The true democrat deifies his nation just as little as the religious person in our sense does."

Einstein's conception of the relation between religion and science is very similar to that prevailing in the liberal Protestant churches of America. As an example one need only cite the views of an outstanding representative of American science such as Robert Millikan. According to this conception, science can never be criticized or directed by religion, since it deals with very different aspects of human life. Millikan once said:

"Let me show why in the nature of things there can be no conflict. This appears as soon as one attempts to define what is the place of religion in human life. The purpose of science is to develop without prejudice a knowledge of the facts and the laws of nature. The even more important task of religion, on the other hand, is to develop the conscience, the ideals and the aspirations of mankind."

This conception of religion abandons completely any demand for belief in any specific scientific or historical facts, and regards religion as a social institution, the purpose of which is to promote a certain attitude toward life and a certain type of behavior in our daily living. Einstein's conception of religion fits in very well with this general attitude. Consequently we can understand why English and American clergymen in particular have been so much interested in Einstein.

6. *Beginning of the Atomic Age*

The dramatic climax with which the second World War was brought to an end by the atomic bomb again brought Einstein's name to public attention. The result that he had derived from his special theory of relativity in 1905 — namely, that mass and energy are equivalent — was demonstrated to the world with almost incredible force of destruction.

As was mentioned in Section 7 of Chapter III, there are nuclear transformations in which a part of the mass of the atomic nucleus is transformed into energy. Numerous such reactions were discovered by scientists, but in all cases the energy required to perform the transformation was much greater than that obtained from the reaction. Hence the practical use of nuclear transformations as a source of power did not seem feasible.

The whole picture changed, however, with the discovery by Otto Hahn and Lise Meitner of the fission of uranium. These scientists at the Kaiser Wilhelm Institute in Berlin found that when uranium is bombarded with neutrons, its nucleus sometimes breaks up into two more or less equal parts with the liberation of a tremendous quantity of energy. When this news was communicated to other laboratories, the startling result was immediately confirmed. Furthermore, Enrico Fermi, an Italian physicist who had fled to the United States from the Fascist regime, pointed out the possibility, which was soon found to be true, that this breaking up of the uranium nucleus is accompanied by the production of several neutrons. The important significance of this last discovery lies in the fact that this process, named the "fission" of uranium, may be made self-sustaining. Once the process is started, the neutrons produced by the fission of one uranium nucleus can cause others to break up, and the neutrons from these can, in turn, cause other fissions. Thus a self-perpetuating nuclear "chain reaction," in which a large number of nuclei breaks up with the consequent liberation of a tremendous amount of energy, became a possibility. Calculations showed that as much energy would be released by the fission of a pound of uranium as by the burning of thousands of tons of coal.

It soon became evident to many scientists that this liberation of energy may be made to take place almost instantaneously, and that consequently uranium bombs with millions of times the destructive power of ordinary explosives could be produced. It was also evident to them that if such an instrument came into the hands of the fascist nations, they would use it in their war of aggression, and civilization would then be doomed. Such apprehensions were felt especially strongly by those scientists who had fled from the persecutions in their native countries. Two physicists at Columbia University, a Hungarian named Leo Szilard, who had fled from the University of Berlin, and the aforementioned Fermi, became convinced that the military authorities in the United States ought to be informed about this possibility. Moreover, Szilard realized that unless this problem was taken to a government official in a very high position, their words would not be heeded. He had been acquainted with Einstein in Berlin and it seemed to Szilard that Einstein's great reputation and world-wide recognition as a physicist could be used to convince the authorities of the importance of this problem. He therefore got in contact with Eugene Wigner, another Hungarian physicist, then teaching at Princeton University, and in July 1939 they conferred with Einstein.

At that time the average engineer, civilian or military, regarded the theory of relativity as something very bookish, which only impractical college professors talked about and which would never have any industrial application. And as for nuclear physics, he had not even heard of it. It was therefore obvious that the problem of interesting the government in the practical use of atomic energy and of obtaining funds for its development was a difficult one. To these scientists it seemed that if anybody would respond to such a suggestion, it was President Roosevelt. He had been aware of the Nazi policy of aggression from the very beginning, and he was fully cognizant of the threat to the future security of this country. Moreover, he was not as firmly convinced of the foolishness of college professors as most politicians are.

In view of these circumstances, Szilard and Fermi suggested to Einstein that he make a direct appeal to the President.

As we have already seen, Einstein disliked becoming involved in public affairs, and he felt a special reluctance to give advice on military affairs and to encourage the development of the most devastating weapon yet discovered by man. On the other hand, he was convinced that the Nazis would be in possession of atomic power in the near future and would use it to subdue the rest of the world. With the responsibility that he felt in his exceptional position as the most famous scientist in this country, he realized what his duty was.

On August 2, 1939 Einstein addressed a letter to President Roosevelt, which began:

"Some recent work by E. Fermi and L. Szilard which has been communicated to me in manuscript leads me to expect that the element Uranium may be turned into a new and important source of energy in the immediate future. . . . A single bomb of this type . . . exploded in a port . . . might very well destroy the whole port, together with the surrounding territory. . . ."

Furthermore, Einstein advised the President of the probability that research in this field might be far advanced in Germany, and stressed the great danger that the United States would incur if the Nazis got hold of such a bomb. Einstein suggested that a special organization with a staff of scientists who had devoted themselves to nuclear research should be created to carry on the investigations on the practical use of uranium.

The result of this project which was so dramatically made public and the subsequent publicity concerning the organization and development of the Manhattan Project, as it was later called, are too well known by now to need reiteration here.

The immediate reaction of the American people to the announcement of the atomic bomb and the Japanese surrender, which soon followed, was the feeling of relief that the war was over and of pride that the United States had proved to be in the lead in science.

The scientists who had worked on the development of the atomic bomb, however, saw in it a political implication that gave them cause for alarm. The war had been brought to an end with a brilliant victory for the democracies, but the establishment of peace seemed to lead to an impasse. An atmosphere of distrust had arisen among the Allies which could easily sow the seed for another war. Moreover, the atomic bomb now made it possible for an aggressor nation to make a surprise attack that would practically annihilate its opponent within a few minutes. The scientists felt the full weight of the responsibility that they had created, and they began to take action in educating the Congress and the public in general. They wanted the whole nation to realize the full gravity of the situation. The "secret" of the atomic bomb would be shortlived and there is no adequate defense against it.

For Einstein, who had been instrumental both in the development of the basic theory and in the approach to President Roosevelt, the responsibility weighed doubly hard. He agreed wholeheartedly with the scientists like Oppenheimer and Shapley who tried their best to explain the full implications of the new weapon to the politicians and the military authorities. Einstein, however, has always disliked getting involved in politics and he was never willing to compromise his views with the troubles of the next day. He is in full agreement with the view expressed by Emery Reves in his book *The Anatomy of Peace,* in which we read: "We must grasp the fact that it is necessary to limit the sovereignty of nations and to establish a world government which will regulate the relations between nations by law, as the United States, for example, now regulates the relations between states." For this reason Einstein is not satisfied with the suggestion to hand over the secret of the atomic bomb to the principal members of the United Nations or even to the United Nations organization itself.

Since no world government exists at present, however, Einstein's view seems to suggest that the secret should remain for the time being with the original manufacturers, the United States, Great Britain, and Canada. Hence, he was accused by some people of being idealistic and impractical and by others of being reactionary and taking sides with the "brass hats."

When I discussed with Einstein recently his views on the international aspects of the atomic bomb he protested vehemently against these interpretations of his views. He realizes exactly that the "control of atomic energy" is primarily not a technical but a political problem which cannot be solved except in the form of a peace settlement between the big nations. Every "control" requires an international agreement that entrusts agents with the supervision of the war research and industry of all nations. Such an agreement presupposes a high degree of mutual trust, and if such a trust exists, there will be no danger of war, bomb or no bomb.

Einstein realizes that this vicious circle cannot be broken by singling out the "control of atomic energy" but only by a comprehensive territorial and economic agreement. He hopes that the fear of atomic warfare may become so great that governments and people will be prepared to sacrifice their sovereignty to a greater degree than they would without this threat.

7. *Life in Princeton*

Einstein's wife Elsa died in 1936. She had been strongly attached emotionally to her German homeland, and after losing her, Einstein became even more strongly linked to his new country. His first wife never left Switzerland, but their eldest son, who had been born in Bern at the time of Einstein's first great discoveries, is now also active in the United States as an engineer. Of Einstein's two stepdaughters, one died after leaving Germany; the other, Margot, a talented scupltress, was divorced from her husband and now lives mostly with Einstein in Princeton.

In 1939 Einstein's only sister Maja moved from Florence, Italy, to Princeton. She is married to the son of that teacher Winteler of the cantonal school of Aarau, to whom Einstein had been strongly attracted. She had felt uneasy because of the increasing Nazi influences in Italy. Her husband returned to Switzerland temporarily, while she visited her brother. Her manner of speaking and the sound of her voice, as well as the childlike and yet skeptical formulation of every statement, are unusually similar to her brother's mode of expression. It is amazing to listen to her; it arouses a sense of uneasiness to find a replica of even the minor traits of a genius. Nevertheless, there is also a certain feeling of reassurance at seeing even the greatest genius as a link in a chain of ordinary natural events.

Since 1928 Einstein's secretary and later his housekeeper is Miss Helen Dukas. She is trim, intelligent, and energetic. She is a native of Einstein's Swabian homeland, and comes from the same small place as Elsa Einstein. These three women now form Einstein's immediate environment.

In 1933 when Einstein came to the United States he had only a visitor's visa. Under the American immigration law there is no place within the country where one can obtain permission to become a permanent resident of the United States. Such permission can be given only by an American consul, and these officers are only to be found in foreign countries. Consequently Einstein went to the English colony of Bermuda in order to apply to the American consul there. Einstein's visit to the island was a gala occasion. The consul gave a dinner in his honor and gave him permission to enter the United States as a permanent resident.

Only then was Einstein able to announce his intention of becoming a citizen of the United States and to receive his first citizenship papers. He still had to wait five years before he could become a citizen. During this time he had to prepare himself for an examination on the American Constitution and the rights and duties of an American citizen. This he did with zeal. In 1941 Einstein together with his stepdaughter Margot and his secretary, Miss Dukas, received American citizenship. He was asked to broadcast to the public the ideas and emotions which he felt at that moment.

Thus this mighty tree with its roots was transplanted into new earth. What is his life here?

Various things from his Berlin apartment have been brought to his cottage, situated in the midst of a large garden on a suburban street. Here various rare objects such as adorned the living-room of a well-to-do Berlin family can be seen again, for example, Byzantine icons from Russia with their gold background mysteriously darkened by incense. At Princeton Einstein actually lives like a strange guest just as he did in the upper middle-class household in Berlin. His profoundly bohemian nature has not changed even with his sixtieth birthday, which he celebrated at Princeton in 1939.

He has no social life in the traditional meaning of the word. He takes no part in the series of dinners and receptions that are given by the faculty members in university communities. The conclusion should not be drawn, however, that he does not like to see people. On the contrary, he likes to receive people whom he can advise or help, with whom he can discuss some interesting subject or have a pleasant chat, or, what he prefers most of all, with whom he can play music together. He likes people who are ready and enthusiastic to accompany his violin playing on the bass viol, or cello, or the piano. Most of his visitors are not members of Princeton University or of the Institute for Advanced Study. His thoughts are always more occupied with things that are distant than with those that are near. Nevertheless, hardly an afternoon passes without a visitor from out of town coming to talk with him.

Among these visitors are, first of all, physicists, philosophers, or even theologians who come to Princeton and want to use the occasion to obtain some impression of the man who has given their particular field so many new ideas. There is also the great number of refugees from Europe who seek advice and help from him. Sometimes there are people from Europe who stay with him for a few days because they are destitute. There are Zionists who want to hear his position on certain political questions. Even members of the faculty of the University of Jerusalem come to him because they want him to intervene in their favor. There are writers, journalists, artists, who want to interest him in their work, hoping thereby to find a larger audience. The number of people who wish to see him is great, and Miss Dukas has to use a good deal of tact, energy and kindness to keep the atmosphere around him as quiet as he needs it.

His attitude in this matter is the same as in all problems of social life; he feels himself very much apart from other people and he can never identify himself very strongly with others. He always has a certain feeling of being a stranger, and even a desire to be isolated. On the other hand, however, he has a great curiosity about everything human and a great sense of humor, with which he is able to derive a certain, perhaps artistic pleasure from everything that is strange and even unpleasant. Finally, he is very good-natured and feels strongly the equality of all human beings. Perhaps he often says to himself: It is just the most unpleasant people with whom one should be least inclined to be short, since they suffer most because no one wants to talk with them.

As a result it is often the rejected inventors and other misunderstood geniuses who come to him. Ever since the time when he was employed in the patent office at Bern he has retained a certain pleasure in listening to the most senseless projects. All of them contain some element of human inventive faculty, even though in a distorted state; and for Einstein's active and penetrating mind it has always been a pleasure to follow through a confused train of thought, to unravel it, and to find the errors in it.

Occasionally he is also visited by physicists who are carrying on research on the basis of ideas that do not agree with those recognized as correct by present-day physicists. Such aberrant scientists can equally well be forerunners of important innovations or simply muddle-headed fellows. Einstein is more willing than others to listen to such physicists and to give careful consideration to their ideas, since it is always a pleasure for him to see the possible seeds of future ideas. At any rate, it is pleasant mental exercise for him to follow through logically a series of deductions, without being sure at the start whether they lead to any reasonable or useful conclusion.

It sometimes happens, however, that some of these inventors

and scientists would feel insulted if he did not accept their conclusions as correct. Just because he is possibly the only famous physicist who is willing to listen to them and to consider their ideas, all the hatred of an unrecognized physicist for those who have achieved fame occasionally concentrated itself against Einstein. This led to the paradoxical result that he was sometimes attacked and condemned most severely by the very persons to whom he had devoted the most attention.

Since immigrating to America, Einstein has rarely spoken at public meetings. Organizations of all kinds have tried to induce him to do so, but he has spoken only when the subject concerned was one in which he was greatly interested. Neither has he attended scientific meetings very often. Only very few times has he discussed his actual researches in professional circles. Nor has he done it very readily, because he often felt that his work was not in line with the trend of research preferred by most physicists. His work has been devoted through many years to the construction of a "unified field theory" that would eventually account for the subatomic phenomena too. He often thought that the researches with which he was occupied would not be received with much interest by those who believed that they should not divert their attention from the central task of today's physics, the interpretation of atomic phenomena by means of Bohr's quantum theory or his principle of complementarity. On some occasions, however, Einstein lectured at scientific congresses on his views on the present and future of physical science in general. One such rare instance was his address in Philadelphia "On Physical Reality," in which the sentence occurs that forms the motto of this book.

The world around Einstein has changed very much since he published his first discoveries. He began his work during the time of the German Kaiser in the environment characteristic of the German and Swiss petty bourgeoisie; he lived during the second World War in the last bulwark of democracy, the United States of America. He was able to make a substantial contribution toward an earlier conclusion of the war than had been expected, and is now anxious to help in making the peace a lasting one. But his attitude to the world around him has not changed. He has remained a bohemian, with a humorous, even seemingly skeptical approach to facts of human life, and at the same time a prophet with the intense pathos of the Biblical tradition. He has remained an individualist who prefers to be unencumbered by social relations, and at the same time a fighter

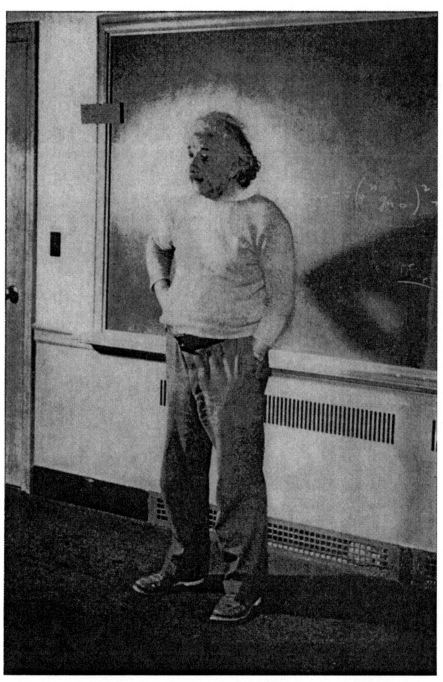

Einstein in his study at the Institute of Advanced Study, Princeton, 1940

LOTTE JACOBI

Einstein in the study of his home 112 Mercer Street, Princeton, 1938

for social equality and human fraternity. He has remained a believer in the possibility of expressing the laws of the universe in simple, even though ingenious mathematical formulæ, but at the same time doubting all ready-made formulæ that claim to be the correct solution for human behavior in private and political life.

When a visitor whom he has known in the old country comes to his home in Princeton, Einstein often says: "You are surprised, aren't you, at the contrast between my fame throughout the world, the fuss over me in the newspapers, and the isolation and quiet in which I live here? I wished for this isolation all my life, and now I have finally achieved it here in Princeton."

Many famous scholars live in the distinguished university town, but no inhabitant will simply number Einstein as one among many other famous people. For the people of Princeton in particular and for the world at large he is not just a great scholar, but rather one of the legendary figures of the twentieth century. Einstein's acts and words are not simply noted and judged as facts; instead each has its symbolic significance — symbolic of his time, his people, and his profession.

People in Princeton tell many anecdotes about Einstein. It is related that one of his neighbors, the mother of a ten-year-old girl, noticed that the child often left the house and went to Einstein's home. The mother wondered at this, whereupon the child said: "I had trouble with my homework in arithmetic. People said that at number 112 there lives a very big mathematician, who is also a very good man. I went to him and asked him to help me with my homework. He was very willing, and explained everything very well. It was easier to understand than when our teacher explained it in school. He said I should come whenever I find a problem too difficult." The girl's mother was alarmed at the child's boldness and went to Einstein to apologize for her daughter's behavior. But Einstein said: "You don't have to excuse yourself. I have certainly learned more from the conversations with the child than she did from me."

I do not know, nor have I made any effort to check, whether this story is true. People tell it in different versions together with the much simpler story that in summer Einstein is often to be seen walking through the streets of Princeton in sandals without stockings, in a sweater without coat, eating an ice-cream cone, to the delight of the students and the amazement of the professors.

Since not only Einstein's personality but also his times and environment should be described in this book, all such stories are certainly true. Even if they do not tell us anything that is factual about Einstein, they are a true description of the world in which he has lived.

In 1945 Einstein retired from his position as professor at the Institute for Advanced Study. This change in his official status, however, did not mean any change in his actual work. He continues to live in Princeton and to carry on research at the Institute.

INDEX

Index

Index